I0055842

Optical
Detection
of
Cancer

Optical Detection of Cancer

editor

Arlen Meyers, MD, MBA

University of Colorado Denver, USA

World Scientific

NEW JERSEY · LONDON · SINGAPORE · BEIJING · SHANGHAI · HONG KONG · TAIPEI · CHENNAI

Published by

World Scientific Publishing Co. Pte. Ltd.

5 Toh Tuck Link, Singapore 596224

USA office: 27 Warren Street, Suite 401-402, Hackensack, NJ 07601

UK office: 57 Shelton Street, Covent Garden, London WC2H 9HE

British Library Cataloguing-in-Publication Data
A catalogue record for this book is available from the British Library.

OPTICAL DETECTION OF CANCER

Copyright © 2012 by World Scientific Publishing Co. Pte. Ltd.

All rights reserved. This book, or parts thereof, may not be reproduced in any form or by any means, electronic or mechanical, including photocopying, recording or any information storage and retrieval system now known or to be invented, without written permission from the Publisher.

For photocopying of material in this volume, please pay a copying fee through the Copyright Clearance Center, Inc., 222 Rosewood Drive, Danvers, MA 01923, USA. In this case permission to photocopy is not required from the publisher.

ISBN-13 978-981-4295-40-6
ISBN-10 981-4295-40-X

Typeset by Stallion Press
Email: enquiries@stallionpress.com

Printed in Singapore.

Preface

Improvements in cancer survival outcomes rest on early detection, elimination of risk factors and better treatment methods. Newer early detection technologies are appearing quickly and include, among others, more sensitive imaging methods, nanodiagnostics, biomarker measurements and, the subject of this book, optical detection devices. The rapid development and deployment of these devices is the result of several factors.

The first driver is the commercial opportunity to exploit bio-photonic technologies that clinically recognize deep and surface — near surface cancers. While imaging and treatment techniques for the management of cancers of the upper aerodigestive system have advanced, for example, the survival outcomes have not changed significantly over the past several decades. There is a compelling need to create affordable, easy-to-use techniques that detect cancer at an early stage in all locations throughout the human body.

Second, recent discoveries in optics, physics and bioengineering has given us a better understanding of the optical, biomechanical, electrical and biomaterials properties of normal and abnormal tissue. As a result, researchers have created innovative ways to detect, amplify and display those differences with high sensitivity and specificity.

Finally, as demonstrated in the contents of this book, the tighter collaboration between global basic researchers, engineers and clinicians has accelerated the rate of biomedical innovation and resulted in faster time to market. New optical biopsy devices are appearing with increasing frequency using a dazzling array of techniques not only to detect cancer, but to monitor the results of treatment, measure the completeness of excision in the operating room in real time, uncover invisible or satellite lesions, and predict the appearance of tobacco related cancers at distant sites.

Thanks to the efforts of the global interdisciplinary researchers and clinicians authors, using commercially available optical detection technologies will soon be the standard of care in most cancer clinics. The result will be less cost, less morbidity and patient suffering, and, hopefully, better outcomes.

I'd like to express my thanks to my colleagues from around the world who have contributed to this volume and tolerated my annoyances in getting it to press.

Arlen D Meyers, MD, MBA

Professor
Otolaryngology, Engineering and Dentistry
University of Colorado Denver
Anschutz Medical Campus
March, 2011

Contents

Chapter 1

The Optical Detection of Cancer: An Introduction

Toby Steele* and Arlen Meyers†

Part I: An Introduction to Optical Detection Technology

Introduction

During 2008, approximately 1.4 million new cases of cancer were diagnosed in the United States, leading to more than 1500 deaths every day.[1] Even with remarkable technological advancements and extraordinary efforts from cancer advocates, scientists, and clinicians, the diagnosis of cancer often occurs at a late stage conferring a dismal prognosis. Importantly, the improvement of patient outcomes is clearly related to the detection of cancerous or precancerous lesions at early stages of disease. Optical detection technology offers the promise to not only detect disease at early stages, but also to improve the monitoring of disease progression or regression during treatment.

The field of optical diagnostics comprises a variety of techniques designed to characterize the relationship between the optical and biological properties of tissue. Through the detection of changes in light after interaction with tissue, optical technologies provide information on the physiologic condition of the tissue at a molecular level. Early research in optical diagnostics suggested that alterations in light-tissue

---

* Department of Otolaryngology, University of California, Davis, USA. E-mail: Tosteele@ucdavis.edu

† Department of Otolaryngology, Dentistry and Engineering, University of Colorado Denver, USA. E-mail: Arlen.Meyers@ucdenver.edu

interactions can be used differentiate normal from malignant tissue.[2] Subsequent advances in molecular biology, genomics, and proteomics, have vastly improved our scientific understanding of the complex biochemical and morphological changes that occur as tissue undergoes the transformation from normal to neoplasia. Many of these early biological events have been shown to alter the optical properties of pre-cancerous and cancerous tissue. Light based detection systems identify these "optical signatures" created during tissue transformation to provide a real-time assessment of tissue structure and metabolism.

Although the pathways responsible for the development of cancer are complex, it is widely accepted that cancer arises through the accumulation of DNA mutations that tip the cell cycle toward proliferation. The proliferation of squamous cancer cells forms a morphologically distinctive spectrum of disease ranging from mild dysplasia to invasive carcinoma.[3] The current identification and diagnosis of precancerous and cancerous lesions relies on the histological and cytological examination performed by a pathologist after suspicious tissue is biopsied. While these methods represent the gold standard for cancer diagnosis, they have several limitations. Tissue biopsy is invasive, expensive, and often time-consuming. The diagnostic interpretation of the tissue sample has been shown to vary amongst pathologists, and the pathologic criteria for the identification of precancerous lesions are not well defined.[4] In addition, early precancerous changes are frequently undetectable by conventional visual inspection, leading to missed opportunities for diagnosis. Optical technologies have the potential to improve these limitations in several ways. Though the benefits provided vary with each technology, optical techniques offering objective data analysis may reduce the variation in pathologic diagnosis. Furthermore, optical technologies show the potential to provide a real-time tissue assessment through a minimally invasive route, eliminating lengthy waits and the need for tissue biopsy. While the benefits of optical technologies are currently limited in clinical practice, the achievement of a highly sensitive and specific optically determined histopathologic diagnosis, an optical biopsy, has the potential to revolutionize medical practice.

Optical detection technology applications

The majority of cancers (~85%) arise in the epithelial tissues that line the interior and exterior surfaces of the human body.[1] Representative cases include cancers of the oral cavity and pharynx, respiratory system, digestive system, genital system, and urinary system. The majority of these cancers are detected visually, generally through the application of endoscopic techniques. Endoscopy involves a fiber based optical device directed by a physician to visualize tissue surface abnormalities. The surveillance and detection of pre-cancerous and cancerous lesions is achieved through images captured by the endoscope, and tissue biopsied after suspicious sites have been identified.[5] The addition of optical technology to conventional endoscopic visualization techniques allows for the identification of lesions often unidentifiable through conventional endoscopy. In addition, the application of optical technology extends beyond the detection of surface cancer. Techniques have been applied to detect cancer in breast and prostatic tissue, along with molecular contrast agents designed to target specific biochemical pathways in the development of cancer.

Techniques for the optical detection of cancer

Spectroscopy

The utilization of spectroscopic techniques for the detection of cancerous and pre-cancerous lesions is based on the analysis of specific light-tissue interactions to assess the state of biological tissue. As tissue undergoes the carcinogenic sequence from normal to neoplasia, complex morphological and molecular transformations occur that modify the manner in which light is absorbed and reflected in the tissue. With the delivery of specific wavelengths of light to tissue through an optical probe, a spectral pattern is collected that contains diagnostic information for tissue classification. Using histologically confirmed tissue specimens from benign and neoplastic tissue, scientists have assembled a spectral database of known light-tissue interactions. The spectra collected from an unknown tissue sample can then be analyzed through various empirical and statistical techniques to produce a histological diagnosis. Spectroscopic techniques such as fluorescence spectroscopy, light scattering spectroscopy and Raman spectroscopy, utilize the unique spectral patterns that are created as tissue progresses towards cancer to offer the potential to detect diseased tissue during the initial stages of carcinogenesis.[7]

Fluorescence spectroscopy

Fluorescence spectroscopy is based on the biological emission of fluorescent light from tissue exposed to ultraviolet (UV) or short wavelength visible (VIS) light. To better understand fluorescence, a brief review of the interaction of light and tissue is warranted. Light is formed by packages of energy termed photons. When tissue is exposed to light, photons may be absorbed, reflected, or scattered by specific molecules within the tissue. As light illuminates the targeted tissue, these biomolecules, termed fluorophores, absorb the energy in the illuminating light and respond by emitting fluorescent light of lower energy (and longer wavelength). The change in wavelength then allows fluorescent light to be differentiated from illuminating light (UV or VIS light). Each group of fluorophores will respond to specific excitation wavelengths, and in turn, emit a different range of wavelengths resulting in a spectral pattern that ideally represents the biochemical and metabolic status of the tissue undergoing optical interrogation.[7-10]

Fluorescent light may be generated by the administration of an exogenous agent such as in drug-induced fluorescence or by the excitation of endogenous fluorophores (autofluorescence). As tissue undergoes the biochemical and morphological progression to neoplasia, the concentration and distribution of the fluorophores is transformed. Known fluorophores include components of the connective matrix (collagen, elasting), metabolic coenzymes (reduced nicotinatimide adenine dinclueotide (NADH), flavin adenine dincucleotide (FAD), flavin mononucleotide (FMN)), aromatic amino acids (tryptophan, tyrosine, phenylalanine), byproducts of the heme biosynthetic pathway (porphryins) and lipopigments (lipofuscin, ceroids). Factors influencing tissue autofluoresence include tissue architecture, light absorption and scattering properties of each tissue layer, the distribution and concentration of the fluorophores in the different tissue layers, the biochemical environment, and the metabolic status of the tissue. Though complex, tissue autofluorescence patterns reflect changes in tissue composition and have been shown to be capable of distinguishing benign from malignant tissue.[7–10,15]

Elastic scattering (reflectance) spectroscopy

Elastic scattering spectroscopy, also known as diffuse reflectance spectroscopy, utilizes the principle of white light (400–700 µm) reflectance to determine the structural characteristics of illuminated tissue. Elastic scattering occurs when photons from visible light are reflected from tissue constituents without a change in wavelength (or energy). The intensity of this back scattered light is measured resulting in a reflectance spectrum that describes the interaction of white light with tissue following multiple scattering events. As tissue transitions to dysplasia or neoplasia, the relative concentrations, density, and size of endogenous scatterers is affected. The measurement of the intensity of back scattered light is then influenced by the characteristics of the scatterers (i.e. nuclei, mitochondria, connective tissue) and absorbers (i.e. hemoglobin). For example, dysplastic change is often characterized by enlarged nuclei, crowding, and hyperchromacity. These changes lead to characteristic reflectance spectra used to identify the structural composition of tissue and aid in clinical diagnosis.[7–10,15]

Raman spectroscopy

Raman spectroscopy is a novel optical technique employed to provide detailed information about the molecular composition of tissue. In contrast to elastic scattering spectroscopy, Raman spectra are generated from the molecule-specific inelastic scattering of light. Following exposure to a light source (generally near-infrared light 700–1300 µm), a minute fraction of the scattered light undergoes a

wavelength shift due to the energy transfer between incident photons and tissue molecules. The wavelength shift (and change in
energy) is achieved when the incident photon alters the vibrational
state of an intramolecular bond. A Raman emission spectrum is generated from the combination of the wavelengths scattered by the
molecules in a tissue sample. These spectral features provide detailed
and specific information about the molecular composition of tissue.
Though Raman spectroscopy is sensitive to a wide range of specific
biomolecules such as proteins, lipids and nucleic acids, the Raman
effect only compromises a small fraction (1 in a million) of scattering
events and signals may be weak and difficult to implement.[8,10,40]
Other modifications of the Raman technique, such as surface
enhancing Raman spectroscopy and coherent anti-stokes Raman
scattering are designed to amplify the signal.

Fluorescence imaging

Fluorescent imaging systems utilize spectroscopic principles to capture fluorescence emission spectra from a larger tissue sampling area
than is possible with point spectroscopy. The acquisition of an image
requires tissue illumination with a light source, often in the near-UV
to green range. The subsequent fluorescence produced from the
absorption and scattering events is recorded with a camera and results
in a real-time image. In addition, fluorescence imaging systems are
capable of sampling larger tissue areas and provide two-dimensional
information allowing for the detection of lesion-specific features such
as homogeneities.[13]

Optical coherence tomography (OCT)

Optical coherence tomography is an innovative optical imaging
technique designed to provide high resolution (~10–20 μm) cross-
sectional images of microscopic sub-surface tissue structures. As the
optical analogue of high frequency B-scan ultrasonography, a imaging technique that detects back scattered sound waves, OCT images
are generated by measuring the intensity of back scattered light
after tissue is probed with a low-power near-infrared light source

(wavelengths ranging from 750 to 1300 μm). Based on the principle of low-coherence interferometry, OCT is able to provide real-time images at a resolution 10 times greater than endoscopic ultrasound, thereby allowing for the identification of microscopic tissue features such as villi, glands, crypts, lymphatic aggregates, and blood vessels. Despite this high resolution, OCT imaging is limited by a depth of penetration of 1–3 mm.

To obtain images, infrared light is delivered to tissue through an optical probe, typically 2 to 2.4 mm in diameter. Various OCT system designs allow for tissue to be scanned a linear, tranverse, or radial fashion and are easily interfaced with endoscopes, laparoscopes, catheters, and hand-held probes. Initially applied to obtain images in the field of ophthalmology, technological advancements have resulted in the clinical application of OCT imaging in a diverse set of medical specialties, including gastroenterology, dermatology, cardiology and oncology.[11]

Narrow-band imaging (NBI)

Narrow-band imaging is a recently developed optical technique designed to enhance the visualization of microvasculature on the mucosal surface. Developed to improve the quality of endoscopic images, NBI systems limit the depth of light penetration into tissue through red, green, and blue optical interference filters. These three filters divide the visible wavelength ranges into three shorter wavelength bands, while increasing the relative contribution of blue filtered light. The increased contribution of blue light is fundamental to creating a narrow-band image, as blue light corresponds to the peak absorption of hemoglobin. The resulting image demonstrates preferential enhancement of the vascular network of the superficial mucosa. To differentiate normal from dysplastic tissue, the microvascular patterns present in the narrow band image are analyzed. Areas of nondysplastic tissue generally have fine capillary patterns of normal size and distribution; while areas harboring dysplasia demonstrate abnormal capillary patterns with increased size, number, and dilation. The clear visualization of vascular patterns through NBI has been shown to enhance the diagnostic capability of endoscopy and offers promise for the early detection of cancer.[10,25]

Multimodal optical imaging

Advances in bioengineering and the continued refinement of optical detection techniques have led to the development of multimodal optical detection systems. These multimodal devices often function in real-time to provide complementary diagnostic information and wide tissue surveillance capability. The ultimate goal of the optical detection systems centers on the achievement of an "optical biopsy". This achievement would allow clinicians to determine a tissue diagnosis based on *in vivo* optical measurements and would eliminate the need for conventional biopsy and histopathological interpretation. Furthermore, the reliable detection of malignant change through the "optical biopsy," will provide the clinician the ability to immediately determine definitive treatment and optimally improve patient outcomes.

Summary

To briefly review the diagnostic information provided by various optical techniques, fluorescence spectroscopy and Raman spectroscopy offer diagnostic information about the biochemical composition of tissue, while reflectance spectroscopy probes changes in light scattering and absorption to characterize tissue morphology. Optical coherence tomography provides high resolution images of tissue morphology, and NBI offers complementary information about tissue microvasculature.

Part II: The Application of Optical Diagnostic Technology in the Upper Aerodigestive Tract

Introduction

Approximately 281,000 cancers of the upper aerodigestive tract were estimated to be diagnosed in 2008.[1] These sites include the oral cavity, larynx, pharynx, hypopharynx, trachea, bronchus, lung, and esophagus. Importantly, the 5-year survival rates for cancers of the upper aerodigestive tract are dismal, and efforts aimed at the early detection of these cancers have yet to significantly improve clinical outcomes.[1] The promise of optical detection technologies centers on the ability to perform the inexpensive, rapid, and accurate diagnosis of early cancers leading to improved survival rates. As with all new medical technologies, the employment of optical diagnostic technologies in clinical practice must meet or exceed the current standards of care. Recent research in optical diagnostic technology has focused on the translation of these technologies from the laboratory to the clinic. Initial research in optical diagnostics described the properties of normal, dysplastic, and cancerous tissue, and provided a foundation for the *in vivo* diagnosis of suspicious areas. The following segments provide a brief introduction into the current knowledge and clinical application of optical detection systems for the early detection of cancers of the upper aerodigestive tract. Subsequent chapters provide further background and detailed descriptions of each technology. Due to the limited nature of this introduction, a comprehensive review of the applications of each optical technology is excluded and will be discussed in other chapters of the book.

Oral cavity

Cancers of the oral cavity account for approximately 3% of malignancies diagnosed annually in the United States. As with other upper aerodigestive tract cancers, 5-year survival rates for oral cavity cancers decrease significantly with delayed diagnosis (Local: 81.8%, Distant 26.5%).[1] Cancers of the oral cavity are believed to progress from premalignant/precancerous lesions, beginning as hyperplastic tissue, with subsequent increasing levels of dysplasia (mild, moderate, severe) into carcinoma *in situ* (CIS), and finally developing into invasive squamous cell carcinoma.[12] Several optical imaging techniques have shown promise in identifying these lesions, and several systems have gained FDA approval for clinical use.

As the most thoroughly investigated optical techniques for the detection and characterization of oral lesions, autofluorescence spectroscopy and imaging systems have been shown to be capable of distinguishing normal oral mucosa from cancerous lesions. In addition, research suggests that autofluorescence techniques are capable of discriminating between lesion types, though sensitivities and specificities reported by researchers have varied. Research suggests that autofluorescence spectroscopy is exceedingly accurate in distinguishing lesions from healthy oral mucosa (sensitivity 82–100%, specificity 63–100%), though there is a lack of compelling evidence for the discrimination between lesion types.[13] Autofluorescence imaging systems, such as the commercially available VELscope, allow the clinician to probe oral cavity tissue for the direct visualization of precancerous and cancerous lesions. Studies demonstrate that oral cancer and precancerous lesions show a characteristic decrease in green fluorescence when probed with autofluorescent imaging systems. This fluorescent pattern allows for the clinician to visualize malignant changes of oral tissue that manifest as darkened areas surrounded by healthy, green fluorescent tissue.[12]

In addition to fluorescence spectroscopy and imaging techniques, several additional optical diagnostic systems have demonstrated potential for the successful evaluation of the oral cavity. A recent study using a multispectral imaging system (fluorescence, narrow-band imaging, and orthogonal polarized reflectance) demonstrated that oral lesion borders change with each imaging modality, suggesting that multimodal imaging can provide important diagnostic information not available through conventional white light examination or through the use of a single imaging mode alone.[14] Trimodal spectroscopy (Fluorescence spectroscopy, Elastic scattering spectroscopy, Raman spectroscopy) has been shown to be capable of diagnosing malignant/precancerous tissue with a sensitivity and specificity of 96%.[15] Despite the diagnostic advantages created by the combination of optical technologies, these complementary techniques may prove to be time-consuming and expensive, limiting clinical utility.

The application of optical coherence tomography (OCT) for the evaluation of oral cavity disease began as early as 1998 when researchers obtained images of the human tooth and oral mucosa.[16,17] In 2004, OCT images captured from varying states of pathology in hamster cheek mucosa were used to study the feasibility of OCT scanning for oral disease diagnosis.[18] Recent improvements in OCT technology have led researchers to study the clinical utility of OCT for oral cancer diagnosis. Images obtained from *in vivo* benign and malignant oral tissue demonstrated that OCT is capable of recognizing differences in mucosal and submucosal tissue structures allowing for image correlation with known histological features.[19] As OCT technology continues to evolve, faster scanning speeds and higher-resolution images will improve characterization of tissue structure with the hope that this optical modality could improve the detection and management of early stage oral disease.

Pharynx, hypopharynx, and larynx

The upper aerodigestive tract descends from the oral cavity into the pharynx, hypopharynx, and larynx where these anatomical structures contribute to human functions such as swallowing and speech. Similar to the other cancers of the upper aerodigestive tract, pharyngeal, hypopharyngeal, and laryngeal cancers are frequently diagnosed at advanced stages of disease often conferring a dreary prognosis.[1] In particular, pharyngeal and hypopharyngeal cancers are often asymptomatic and difficult to detect until late stages, leading to extensive and sometimes disfiguring surgical and medical interventions. Presently, the detection of these cancers is achieved through endoscopic visualization performed with indirect and direct white-light laryngoscopy, with direct laryngoscopy under general anesthesia and biopsy representing the gold standard for diagnosis. Despite the diagnostic accuracy of laryngoscopy, assessment of precancerous and cancerous lesions may be limited by physician experience, difficult anatomy, and patient discomfort.

Efforts to improve the diagnostic information provided during the evaluation of pharyngeal, hypopharyngeal, and laryngeal surfaces incorporate several of the novel optical diagnostic techniques. Adjunctive optical techniques for the detection of these lesions include autofluorescence laryngoscopy, narrow-band imaging, and optical coherence tomography. Indirect autofluorescence endoscopy has recently been shown to improve routine white-light endoscopy in the evaluation of suspected precancerous and cancerous laryngeal lesions. This imaging technique improved sensitivity by 7% and specificity by 18% and may easily be implemented as an outpatient procedure.[20–22] Additionally, the fluorescence staining of laryngeal neoplasms after the topical application of photosensitizer 5-aminole-vulinic acid (ALA) has been described as a promising diagnostic procedure for differentiating tumor from normal mucosa during microlaryngoscopy.[24] Research investigating the value of narrow band imaging for the early detection of laryngeal cancer demon-strates that NBI can provide high sensitivities and specificities on the basis of abnormal intraepithelial microvasculature changes.[24] Furthermore, research suggests that NBI may improve the sensitivity for the discovery of pharyngeal lesions over conventional methods in addition to serving as an ideal surveillance method after chemoradi-ation therapy for oropharyngeal and hypopharyngeal cancers.[25,26] Optical Coherence Tomography has also shown promise as an imag-ing device for laryngeal cancer, reliably identifying the loss of basement membrane integrity in patients with laryngeal cancer.[27] However, the diagnostic utility of OCT for laryngeal lesions requires further research. Ongoing investigation into the implementation and refinement of these optical techniques demonstrates increasing potential for the role of optical diagnostics in the detection and diag-nosis of pharyngeal, hypopharyngeal, and laryngeal cancers.

Trachea, bronchus, and lung

The application of optical detection modalities in the diagnosis of airway disease is continually evolving. As the most common cause of cancer death worldwide, lung cancer typically presents at a late stage and with a dismal prognosis.[1] The focus of various optical detection techniques (ODT) in the diagnosis of airway disease is in the iden-tification and discrimination of benign (i.e. bronchitis), pre-invasive, and malignant lesions in the central airway.

Autofluorescence bronchoscopy (AFB) has shown considerable promise in the evaluation of pre-invasive airway disease, as evidenced by improved sensitivity compared to white-light bronchoscopy.[28] However, despite this improved sensitivity, screening for pre-invasive lesions with AFB has not been recommended outside controlled clinical trials.[29] In addition to AFB, optical coherence tomography has been shown to be capable of discriminating invasive cancer from CIS, and dysplasia from metaplasia, hyperplasia, and normal tissue, and has the potential to serve as an optical biopsy technique as an adjunct to AFB.[30] The use of Narrow-Band Imaging in the evaluation of airway disease is also under investigation. Research indicates that NBI is capable of improving detection of bronchial dysplasia compared to WLB alone, but requires further evaluation as a stand-alone technology.[31]

Esophagus

The incidence of esophageal cancer has shown a recent dramatic rise in both Europe and the USA, a result mainly attributed to Barrett's esophagus.[32,33] Barrett's esophagus is a condition in which the normal lining of the upper GI tract is replaced by an epithelium that is associated with an increased risk of developing adenocarcinoma.[34] Unfortunately, the five year survival rate for all stages of esophageal cancer is only 15.6%, and improving survival for these cancers must focus on the early detection of cancerous changes.[1] The detection of esophageal cancers is achieved through endoscopic evaluation with biopsy, and several optical techniques have recently been developed to increase the diagnostic accuracy of endoscopic evaluation.

The use of autofluorescence spectroscopy and raman spectroscopy are each promising optical techniques for the detection of upper GI cancers. Data from studies performed to evaluate autofluorescence spectroscopy shows a wide range of sensitivities (75–100%) and specificities (65–95%), and this technique demonstrates promise as an adjunctive optical technique.[35–39] Raman spectroscopy appears capable of identifying high grade dysplasia/early adenocarcinoma with a reported sensitivity of 88%, specificity of 89%, and accuracy of 89%.[40]

Several optical imaging systems have been developed to enhance endoscopic diagnosis of esophageal lesions. Recent studies have shown that autofluorescent imaging systems may increase the detection rate of early esophageal cancer from 63 to 91%, though autofluorescent imaging is often limited by a high false positive rate.[41] Trimodal imaging systems are also under investigation for the optical detection of esophageal cancer. Incorporating high resolution endoscopy, autofluorescence imaging, and narrow band imaging, the trimodal system was able to increase the detection of early neoplasia with the addition of autofluorescence, and reduce the false positive rate with the addition of narrow band imaging.[42]

Optical coherence tomography has also been investigated for the detection of upper GI cancerous changes with promising results.[43–45] A recent study using ultrahigh resolution OCT demonstrated the feasibility of carrying out OCT imaging in conjunction with standard endoscopy for the evaluation of Barrett's esophagus, dysplasia, and adenocarcinoma, though future research will need to focus on evaluating its clinical utility.[46] The recent research and emphasis on enhancing optical techniques for the detection of early esophageal cancerous changes shows significant promise as clinicians seek to alter the poor prognostic implications of cancer diagnosis.

Conclusion

Optical diagnostic technologies utilize the interaction of light and tissue to provide objective data that describes the biological, chemical and morphological composition of the tissue under interrogation. While many of these techniques have only recently been implemented in medical settings, they offer scientists a highly sought after method for the early detection of cancer. Many of the optical diagnostic techniques are still in the research and development stage and before they can be implemented in widespread clinical practice, further research in large clinical trials must confirm the initial experimental data provided by researchers across the globe. With further research and optimization of optical diagnostic technology, the early detection of cancer is an increasingly realistic goal in medical practice.

References

Part I: An Introduction to Optical Detection Technology

1. ACS. Cancer Facts & Figures 2008. American Cancer Society, 2008.
2. Lycette RM, Leslie RB. (1965) Fluorescence of malignant tissue. *Lancet* **40**: 436.
3. Lodish HF, Berk A, Matsudaira P, Kaiser CA, Krieger M, Scott MP, Zipursky SL, Darnell J. (2003) *Molecular Cell Biology*, 5th Ed. W.H. Freeman and Company, New York.
4. Hu XH, Lu JQ. Optical detection of cancers. Encyclopedia of Biomaterials and Bioengineering. Taylor & Francis. 2005.
5. Zara JM, Lingley-Papadopoulos C. (2008) Endoscopic OCT approaches toward cancer diagnosis. *J Sel Quantum Electron* **14**: 70–81.
6. Jemal A, Siegel R, Ward E *et al.* (2007) Cancer Statistics, 2007. *CA Cancer J Clin* **57**: 43–66.
7. Sokolov K, Follen M, Richards-Kortum R. (2002) Optical spectroscopy for detection of neoplasia. *Curr Opin Chem Biol* **6**: 651–658.
8. Crow P, Stone N, Kendall CA *et al.* (2003) Optical diagnostics in urology: Current applications and future prospects. *BJU Int* **92**: 400–407.
9. DaCosta RS, Wilson BC, Marcon NE. (2003) Photodiagnostic techniques for the endoscopic detection of premalignant gastrointestinal lesions. *Dig Endosc* **15**: 153–173.
10. Wong KSLM, Wilson BC. (2005) Endoscopic detection of early upper GI cancers. *Best Pract Res Clin Gastroenterol* **19**: 833–856.
11. Zysk AM, Nguyen FT, Oldenburg AL *et al.* (2007) Optical coherence tomography: A review of clinical development from bench to bedside. *J Biomed Opt* **12**: 051403.

Part II: The Application of Optical Diagnostic Technology in the Upper Aerodigestive Tract

Oral cavity

12. Lane PM, Gilhuly T, Whitehead P *et al.* (2006) Simple device for the direct visualization of oral-cavity tissue fluorescence. *J Biomed Opt* **11**: 024006.
13. De Veld DCG, Witjes MJH, Sterenborg HJCM, Roodenburg JLN. (2005) The status of *in vivo* autofluorescence spectroscopy and imaging for oral oncology. *Oral Oncol* **41**: 117–131.
14. Roblyer D, Richards-Kortum R, Sokolov K, *et al.* (2008) Multispectral optical imaging device for *in vivo* detection of oral neoplasia. *J Biomed Opt* **13**: 024019.
15. Swinson B, Jerjes W, El-Maaytah M *et al.* (2006) Optical techniques in diagnosis of head and neck malignancy. *Oral Oncol* **42**: 221–228.
16. Colston BW Jr., Everett MJ, Da Silva LB, *et al.* (1998) Imaging of hard and soft-tissue structure in the oral cavity by optical coherence tomography. *Appl Opt* **37**: 3582–3585.
17. Colston BW Jr., Everett MG, Sathyam US *et al.* (2000) Imagin of the oral cavity using optical coherence tomography. *Monog Oral Sci* **17**: 32–55.
18. Matheny E, Hanna N, Jung W, *et al.* (2004) Optical coherence tomography of malignancy in hamster cheek pouches. *J Biomed Opt* **9**: 978–981.
19. Ridgway JM, Armstrong WB, Guo S, *et al.* (2006) *In vivo* optical coherence tomography of the human oral cavity and oropharynx. *Arch Otolaryngol Head Neck Surg* **132**: 1074–1081.

Pharynx, hypopharynx, and larynx

20. Arens C, Dreyer T, Glanz H. (2004) Indirect autofluorescence laryngoscopy in the diagnosis of laryngeal carcinoma and its precursor lesions. *Eur Arch Otorhinolaryngol* **261**: 71–76.
21. Arens C, Malzahn K, Dias O, Glanz AM. (1999) Eddoskopische bildgebende Verfahren in der Diagnostik des Kehlkopfkarzinoms und seiner Vorstufen. *Laryngol Rhinol Otol* **78**: 685–691.

22. Malzohn K, Dreyer T, Glanz H, Arens C. (2002) Autofluorescence endoscopy in the diagnosis of early laryngeal cancer, its precursor lesions. *Laryngoscope* **112**: 488–493.
23. Mehlman M, Betz CS, Stepp H *et al.* (1999) Fluorescence staining of laryngeal neoplasms after topical application of 5-aminolevulinic acid: Preliminary results. *Lasers Surg Med* **25**: 414–420.
24. Watanabe A, Masanobu T, Tsujie H, *et al.* (2009) The value of narrow band imaging for early detection of laryngeal cancer. *Eur Arch Otorhinolaryngol* **266**: 1017–1023.
25. Watanabe A, Tsujie H, Taniguchi M, *et al.* (2006) Laryngoscopic detection of pharyngeal carcinoma *in situ* with narrowband imaging. *Laryngoscope* **116**: 650–654.
26. Katada C, Nakayama M, Tanabe S *et al.* (2008) Narrow band imaging for detecting metachronous superficial oropharyngeal and hypopharyngeal squamous cell carcinomas after chemoradiotherapy for head and neck cancers. *Laryngoscope* **118**: 1787–1790.
27. Armstrong WB, Ridgway JM, Vokes DE *et al.* (2006) Optical coherence tomography of laryngeal cancer. *Laryngoscope* **116**: 1107–1113.

Trachea, bronchus, and lung

28. Kennedy TC, Lam S, Hirsch FR. (2001) Review of recent advances in fluorescence bronchoscopy in early localization of central airway lung cancer. *Oncologist* **6**: 257–262.
29. Haeussinger K, Becker H, Stanzel F *et al.* (2005) Autofluorescence bronchoscopy with white light bronchoscopy compared with light bronchoscopy alone for the detection of precancerous lesions: A European randomized controlled multicenter trial. *Thorax* **60**: 496–503.
30. Lam S, Standish B, Baldwin C *et al.* (2008) *In vivo* optical coherence tomography imaging of preinvasive bronchial lesions. *Clin Cancer Res* **14**: 2006–2011.
31. Vincent B, Fraig M, Silvestri G. (2007) A pilot study of narrow-band imaging compared to white light bronchoscopy for evaluation of normal airways and premalignant and malignant airways disease. *Chest* **131**: 1794–1799.

Esophagus

32. Pera M, Cameron A, Trasteck V *et al.* (1993) Increasing incidence of adenocarcinoma of the oesophagus and oesophagogastric junction. *Gastroenterology* **104**: 510–513.

33. Hansson L, Sparen P, Nyren O. (1993) Increasing incidence of both major histological types of oesophageal carcinomas among men in Sweden. *Int J Cancer* **54**: 402–407.

34. Cameron AJ, Ott BJ, Payne WS. (1985) The incidence of adenocarcinoma in columnar-lined (Barrett's) esophagus. *N Engl J Med* **313**: 857–859.

35. Georgakoudi I, Jacobson BC, Van Dam J *et al.* (2001) Fluorescence, reflectance, and light-scattering spectroscopy for evaluating dysplasia in patients with Barrett's esophagus. *Gastroenterology* **120**: 1620–1629.

36. Panjehpour M, Overholt BF, Vo-Dinh T *et al.* (1996) Endoscopic fluorescence detection of high-grade dysplasia in Barrett's esophagus. *Gastroenterology* **111**: 93–101.

37. Bourg-Heckly G, Blais J, Padilla JJ *et al.* (2000) Endoscopic ultraviolet-induced autofluorescence spectroscopy of the esophagus: Tissue characterization and potential for early cancer diagnosis. *Endoscopy* **32**: 756–765.

38. Wang KK, Buttar NS, Wong Kee Song LM *et al.* (2001) The use of an optical biopsy system in Barrett's esophagus. *Gastroenterology* **120**: A413.

39. Pfefer TJ, Paithankar DY, Poneros JM *et al.* (2003) Temporally and spectrally resolved fluorescence spectroscopy for the detection of high grade dysplasia in Barrett's esophagus. *Lasers Surg Med* **32**: 10–16.

40. Wong KSLM, Molckovsky A, Wang KK *et al.* (2005) Diagnostic potential of Raman spectroscopy in Barrett's esophagus. *Proc SPIE* **5692**: 140–146.

41. Kara MA, Peters FP, ten Kate FJW *et al.* (2005) Endoscopic video autofluorescence imaging may improve the detection of early neoplasia in patients with Barrett's esophagus. *Gastrointest Endosc* **61**: 679–685.

42. Curver WL, Singh R, Wong Kee Song LM *et al.* (2008) Endoscopic trimodal imaging for detection of early neoplasia in Barrett's oesophagus: A multi-centre feasibility study using high-resolution endoscopy, autofluorescnece imaging and narrow band imaging incorporated in one endoscopy system. *Gut* **57**: 167–172.

43. Evans JA, Poneros JM, Bouma BE *et al.* (2006) Optical coherence tomography to identify intramucosal carcinoma and high-grade dysplasia in Barrett's esophagus. *Clin Gastroenterol Hepatol* **4**: 38–43.

44. Isenberg G, Sivak MV, Chak A *et al.* (2005) Accuracy of endoscopic optical coherence tomography in the detection of dysplasia in Barrett's esophagus: A prospective, double-blinded study. *Gastrointest Endosc* **62**: 825–831.

45. Qi X, Sivak MV, Isenberg G *et al.* (2006) Computer-aided diagnosis of dysplasia in Barrett's esophagus using endoscopic optical coherence tomography. *J Biomed Optics* **11**: 044010.

46. Chen Y, Aguirre AD, Hsiung PL *et al.* (2007) Ultrahigh resolution optical coherence tomography of Barrett's esophagus: Preliminary descriptive clinical study correlating images with histology. *Endoscopy* **39**: 599–605.

Chapter 2

Optical Coherence Tomography in Oral Cancer

Shahareh Sabet* and Petra Wilder-Smith*

Introduction

A person dies from oral cancer every hour of every day in the United States alone.[1] It is estimated that 35,720 men and women in the US will be newly diagnosed with oral cancer in 2009, with a 5-year survival rate of 60%.[2] This cancer has a higher death rate then cervical cancer, Hodgkin's lymphoma, laryngeal cancer, cancer of the testes, endocrine system cancers such as thyroid, or skin cancer (malignant melanoma).[1] The high death rate is attributed to its predominant detection at an advanced stage: 2/3 of all lesions are at an advanced stage at the time of diagnosis. Five-year survival rate is 75% for those with localized disease at diagnosis, but only 16% for those with cancer metastasis.[3] Thus there exists an urgent need to expand our approach to diagnosing oral cancer and to explore new means of early detection, monitoring, and screening modalities.[2,4,5]

Squamous cell carcinoma (SCC) accounts for 96% of all oral cancers. Premalignant changes usually present in the form of white (leukoplakia), red (erythroplakia), or mixed red and white (erythroleukoplakia) lesions on the oral mucosa. Although white lesions are more common, developing in 1–4% of the population,

* Beckman Laser Institute, University of California, Irvine, USA

red lesions have a greater potential for becoming cancerous, with a risk for malignant conversion of 90%.[5,6] Malignant transformation, which is quite unpredictable, develops in 1–40% of leukoplakias over five years.[5,6] A non-invasive diagnostic modality should permit *in vivo* non-invasive high-resolution imaging of epithelial and subepithelial structures, informing on lesion[1] depth and thickness,[2] histopathological appearance, and peripheral margins[3] to delineate treatment needs at an early, relatively harmless stage.

In this chapter we provide a brief overview of existing clinical practice, then we explore the potential of novel diagnostic approaches, with specific emphasis on the current status of Optical Coherence Tomography (OCT) as a diagnostic tool for oral dysplasia and malignancy.

Optical coherence tomography

Optical Coherence Tomography (OCT) was first introduced as an imaging technique in biological systems in 1991.[7] The non-invasive nature of this imaging modality coupled with (i) a penetration depth of 2–3 mm, (ii) high resolution (5–15 μm), real-time image viewing and (iii) the capability for cross-sectional as well as 3-D tomographic images are excellent prerequisites for *in vivo* oral use and the potential diagnosis of precancer and cancer.

OCT has most often been compared to ultrasound imaging. Both technologies employ backscattered signals reflected from different layers within the tissue to reconstruct structural images, with the latter measuring sound rather than light. The engineering principles behind OCT have been described previously.[7,8] OCT is based on low-coherence interferometry using broadband laser light waves emitted from a source that is directed toward a beam splitter and divided into two pathways (Fig. 1). Light traveling down one pathway is directed toward the tissue sample and the other toward a reference mirror of known path length. Reflected beams from the tissue sample and reference mirror are

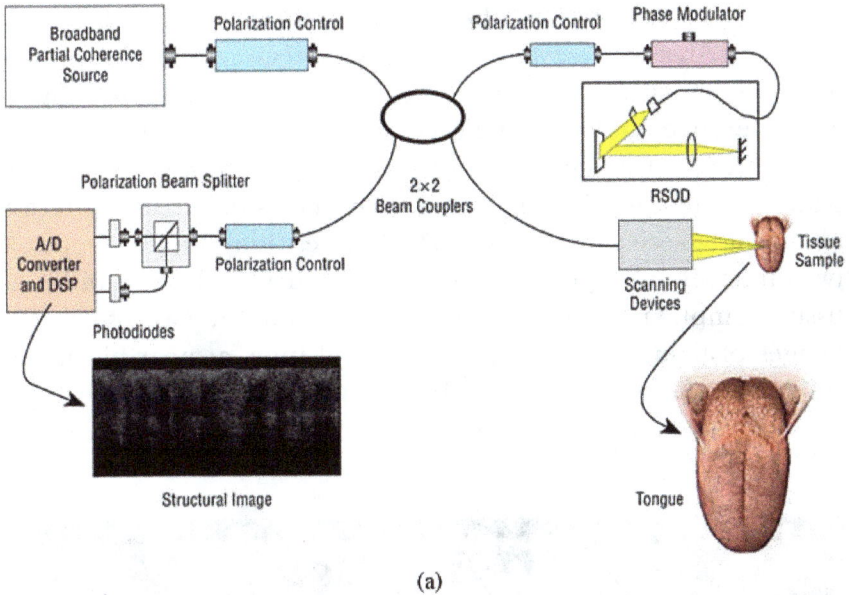

(a)

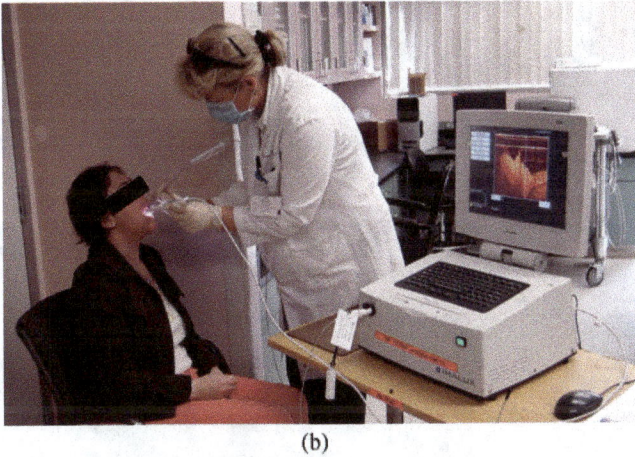

(b)

Fig. 1. (a) Schematic of optical coherence tomographic (OCT) imaging system and oral cavity/oropharynx target. Reprinted from Ridgway, J.M. *et al.* (2006) *Arch Otolaryngol Head Neck Surg,* **132**: 1074–1081. (b) Intra-oral imaging using the Imalux™ OCT system.

directed back towards the beam splitter where they are recombined and detected by a photodetector. Analyses of the interference signal created between the light from the sample and reference mirror can be used to image features of the tissue. Cross sectional images of the tissue sample are constructed by scanning the optical beam transversely across the area of interest obtaining a series of axial signals. The resulting OCT image is a two-dimensional representation of the optical reflection within a tissue sample in cross-section (Figs. 2 and 3). Cross-sectional images of tissues are constructed in real time, at near histologic resolution (approximately 5–15 μm with current technology). These images can be stacked to generate 3-D reconstructions of

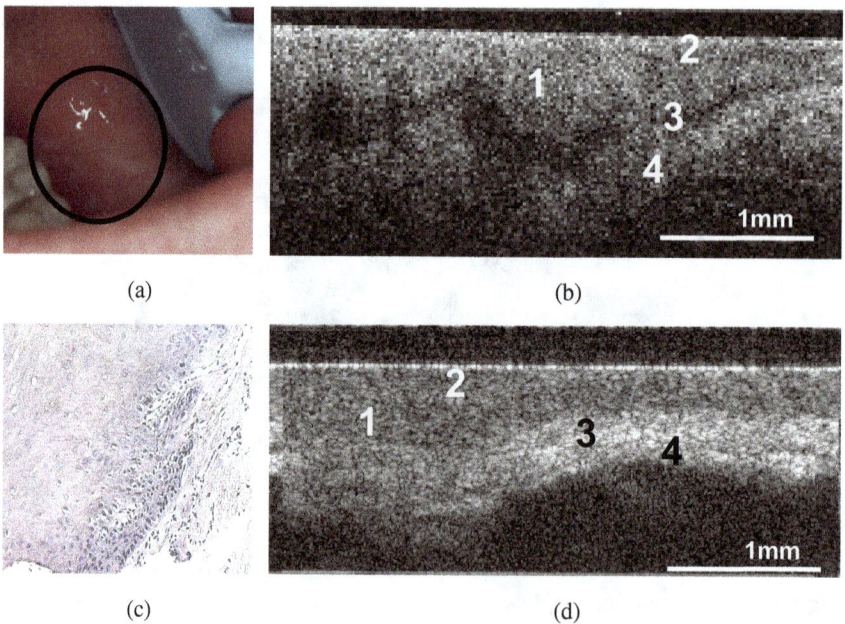

| (a) | (b) |
| (c) | (d) |

Fig. 2. Human buccal mucosa with leukoplakia (white patch). (a) Photograph. (b) *In vivo* OCT image of area of dysplasia. (c) H&E (10×) of dysplastic buccal mucosa. (d) *In vivo* OCT image of normal buccal mucosa. 1 — stratified squamous epithelim, 2 — keratinized epithelilal surface layer, 3 — basement membrane, 4 — submucosa.

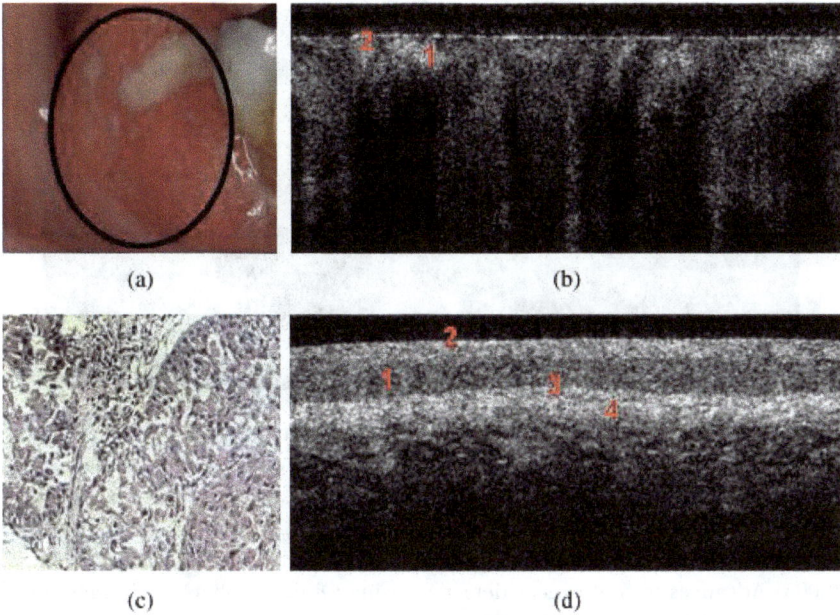

Fig. 3. Squamous cell carcinoma of the buccal mucosa. (a) Photograph. (b) *In vivo* OCT image. (c) H&E (10×) of buccal mucosa with squamous cell carcinoma. (d) *In vivo* OCT image of normal buccal mucosa. 1 — squamous epithelium, 2 — keratinized epithelilal surface layer, 3 — basement membrane, 4 — submucosa.

the target tissue (Fig. 4). This permits *in vivo* non-invasive imaging of epithelial and subepithelial structures, including: (1) depth and thickness, (2) histopathological appearance, and (3) peripheral margins.

Several OCT systems have received FDA clearance for clinical use, and indeed OCT is deemed by many to be an essential imaging modality in field the field of ophthalmology. *In vivo* image acquisition is facilitated through the use of a flexible fiberoptic OCT probe (Fig. 5). The probe is simply placed on the surface of the tissue, to generate real-time, immediate surface and subsurface images of tissue microanatomy whilst avoiding the discomfort, delay and expense of biopsies.

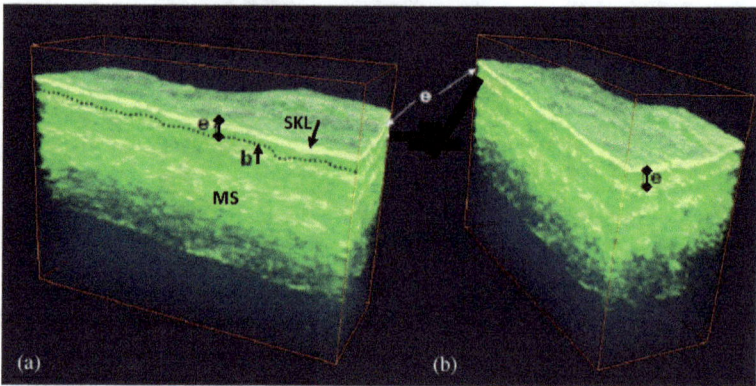

Fig. 4. 3-D reconstructed OCT images of healthy hamster cheek pouch mucosa. The surface squamous keratinized epithelium, and underlying submucosa and muscle layer are clearly visible. (a) Anterior view. (b) Lateral view e — epithelium; b — basement membrane; SKL — surface keratinized layer; MS — muscle. From: Jung WG, Zhang J, Chung JR, Wilder-Smith P, Brenner M, Nelson JS, Chen Z. (2005) Advances in oral cancer detection using Optical Coherence Tomography. *J Biophotonics STQE.* **11**(4): 811–817.

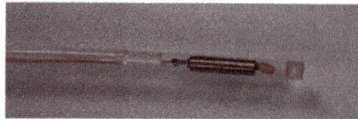

(a)

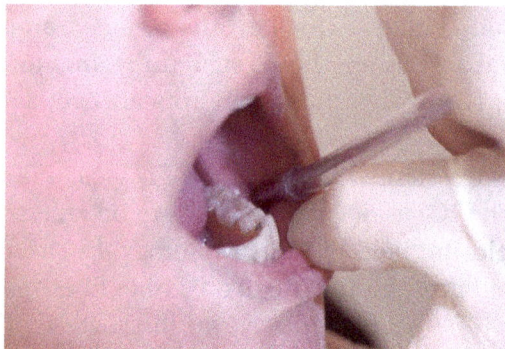

(b)

Fig. 5. (a) Photograph of intra-oral fiberoptic probe. Reproduced with permission from Jun Zhang and Zhongping Chen. (b) *In vivo* intra-oral maging using the Imalux™ OCT probe.

Existing diagnostic tools for oral cancer detection

Currently, visual examination followed by surgical biopsy constitutes the state-of-the-art diagnostic tool for oral pre-cancer and cancer diagnosis.

1. *Visual examination and biopsy*

Primary oral cancer screening is performed by dentists and physicians by means of a thorough oral examination, easily completed in less than five minutes. However, studies show that currently less than 15% of those who visit a dentist regularly report having had an oral cancer screening.[1] Early recognition of malignancy is impeded by the frequent lack of early gross signs or obvious symptoms. Moreover, adequate visual identification and biopsy of all such lesions to ensure that they are all recognized and diagnosed is difficult. Therefore, visual examination alone has poor diagnostic accuracy, providing primarily an indication of the need for tissue biopsy.[1,9–11] Because of the often heterogeneous histopathology of oral lesions, diagnosis based on biopsy at one tissue site may not reflect the true level of risk posed by an entire lesion. The often multi-focal nature of oral malignancy poses added challenges to the biopsy process.[10]

To detect the transformation of leukoplakia/erythroplakia to squamous cell carcinoma, the gold standard currently in practice is regular surveillance accompanied by biopsy or surgical excision. It is impractical to biopsy regularly every oral lesion, and the invasive nature of surgical sampling discourages future patient cooperation and prevents regular screening of high risk sectors of the population.

2. *Oral brush cytology*

The brush biopsy (CDx) is intended for use on clinical lesions in patients who are asymptomatic or those with minor symptoms that are not sufficiently suspicious to warrant biopsy. Cytological examination of "brush biopsy" samples used to detect oral epithelial

dysplasia or squamous cell carcinoma reveal a moderate sensitivity
level (70–100%). However, the specificity of this approach is questionable as studies have reported levels as low as 3% and as high as
100%, necessitating the need for surgical biopsy to support their
diagnostic value.[4,12]

3. *Vital staining*

The sensitivity of staining agents such as Lugol iodine and Tolonium
chloride (Toluidine blue) for the detection of oral malignancy
approximates 90% when used by an expert. However, lower sensitivities are reported when staining is performed by non-experts, or in
the presence of oral inflammation, scarring etc. Reported specificity
is also variable, depending to a large extent on user experience.[4,13–15]

4. *Chemiluminensce: ViziLite*

This screening tool has been FDA-cleared for use in the oral cavity
since November 2001. After rinsing with an acetic acid solution, the
oral cavity is examined under chemiluminescent light (ViziLite)
that generates a moderately short wavelength light with peak out-
puts near 430, 540 and 580 nm. This method allows for visual
distinctions between normal and abnormal epithelium. Abnormal
squamous epithelial cells, typically containing a greater nuclear content density and mitochondrial matrix, preferentially reflect light
and appear distinctly white when viewed under a diffuse low-energy
wavelength light. Normal epithelium will absorb the light and
appear dark.[16] The majority of studies investigating chemiluminescence evaluate subjective perceptions of characteristics of intra-
oral lesions including brightness, sharpness and texture vs. routine
clinical examination. Results have been contradictory.[4,17] Recently a
combination of both Toluidine Blue and Vizilite systems (ViziLite
Plus with TBlue system) has been introduced. A new chemiluminescence device (MicroLux DL) has also recently been introduced into
the market.[18]

5. *Spectroscopy and autofluorescence*

During the disease process, the structural and/or biochemical properties of oral tissue may be significantly altered; spectroscopy can provide information about these properties thus conveying diagnostic information to the examiner. Currently on the market and available to dental practitioners is a tissue fluorescence imaging system called VELscope®. Under illumination by a beam of blue light, healthy oral mucosa emits a pale green autofluorescence while abnormal tissue displays a loss of fluorescence and thus appears darker than the surrounding healthy tissue.[19,20] In a recent study using 56 patients with oral lesions and 11 normal volunteers, normal tissue could be discriminated from dysplasia and invasive cancer with 95.9% sensitivity and 96.2% specificity in the training set, and with 100% sensitivity and 91.4% specificity in the validation set. Disease probability maps qualitatively agreed with both clinical impression and histology.[20] Further clinical studies are needed in diverse populations to fully evaluate the clinical usefulness of this promising approach.

Other devices under development that use a range of spectroscopic techniques, often combined with other approaches include the FastEEM4® System, the Indentafi® and the PS2-oral®. These clinical studies are still at a relatively early stage, and preliminary results are encouraging.[20–27] Remicalm's Identafi™ technology combines anatomical imaging with fluorescence, fiber optics and confocal microscopy with the goal of precisely mapping the location and determining the extent of the disease in the area being screened. A study in 124 subjects determined a sensitivity of 82% and a specificity of 87% for differentiating between neoplastic and non-neoplastic sites in the oral cavity. Results differed between different sampling depths, and keratinized vs. non-keratinized tissues.[25]

Challenges to the use of diagnostic spectroscopy include the often low signal-to-noise ratio, difficulty in identifying the precise source of signals, data quantification issues, and establishing definitive diagnostic milestones and endpoints, especially given the wide range of tissue types contained within the oral cavity. Limited tissue penetration and concerns about mutagenicity when using UV light present further clinical challenges.

6. _In vivo confocal imaging_

This imaging modality resembles histological tissue evaluation in concept. However, three- dimensional subcellular resolution is achieved non-invasively and without the need for staining. In epithelial structures, resolutions of 1 μm have been achieved with a 200–400 μm field of view.[28–30] While this technology can provide detailed images of tissue architecture and cellular morphology, a very small field of view and limited penetration depth reduce the clinical usefulness of this approach.

7. _Photosensitizers_

When topical or systemic photosensitizers are administered, their tendency to accumulate in cancer tissues can be used to identify and delineate areas of pathology. This approach provides for the ability to map in 3-D surface and subsurface lesion margins, inspect large surface areas as well as the potential for subsequent photodestruction of the photosensitized lesion. Many photosensitizing agents are being studied; however FDA clearance for photosensitizing drugs remains limited. Some promising agents for Photodetection include Aminolevulinic Acid (ALA) (Levulan®), Hexyl aminolevulinate (Hexvix®), Methyl aminolevulinate (Metvix®), Tetra(meta-hydroxy phenyl)chlorin (mTHPC®), as well as Porfimer sodium (Photofrin®).[31,32] In a blinded clinical study of 20 patients with oral neoplasms (Fig. 6), the diagnostic sensitivity using unaided visual fluorescence diagnosis or fluorescence microscopy approximated 93%. Diagnostic specificity was 95% for visual diagnosis, improving to 97% using fluorescence microscopy.[31]

Depending on the photosensitizer used and its mode of application (systemic vs. topical), limitations include systemic photosensitization over prolonged periods of time, penetration-related issues, the need for specialized fluorescence detection and

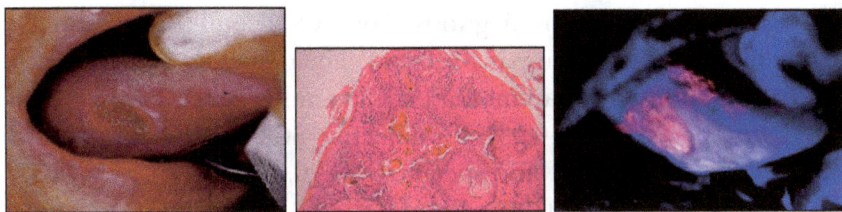

Fig. 6. Photograph, histology and *in vivo* fluorescence images of a tongue with multiple squamous cell carcinoma lesions. Photfrin^R-induced red fluorescence is clearly apparent in multiple lesions of the tongue. From: Chang CJ, Wilder-Smith P. (2005) Topical application of photofrin for photodynamic diagnosis of oral neoplasms. *Plast Reconstr Surg* **115**(7): 1877–1886.

mapping equipment, and lack of specificity when inflammation or scar tissue are present.

OCT Use in the Detection of Oral Cancer

Several studies have sought to investigate the clinical diagnostic capability of *in vivo* OCT to detect and diagnose oral premalignancy and malignancy.[33–35] In a recent independent, blinded study utilizing 50 patients with oral leukoplakia or erythroplakia lesions the effectiveness of OCT for detecting oral dysplasia and malignancy was evaluated.[33] In OCT images of dysplastic lesions epithelial thickening, loss of stratification in lower epithelial strata, epithelial downgrowth and loss of epithelial stratification were visible (Fig. 2). Areas of Oral Squamous Cell Carcinoma of the buccal mucosa were identified in the OCT images by the absence or fragmentation of the basement membrane, as well as an epithelial layer that was highly variable in thickness, with areas of erosion and extensive downgrowth and invasion into the subepithelial layers (Fig. 3). Statistical analysis of the data gathered in this study substantiated the facility of the use of *in vivo* OCT to detect and diagnose oral premalignancy and malignancy in human oral mucosa, with excellent (a) intraobserver agreement between OCT-based diagnosis (Cohen's kappa of 0.872), (b) interobserver agreement for OCT-based diagnosis (Cohen's kappas of 0.870),

(c) agreement between diagnosis from OCT and histopathology respectively (Cohen's kappas of 0.896). For detecting Carcinoma *in situ* or Squamous Cell Carcinoma (SCC) vs. non-cancer, sensitivity was 0.931 and specificity was 0.931; for detecting SCC vs. all other pathologies, sensitivity was 0.931 and specificity was 0.973.

In another study of 97 patients, the ability of OCT to detect neoplasia *in vivo* in the oral cavity was evaluated.[34] The results revealed that the main diagnostic criterion in the OCT image for high-grade dysplasia/cancer *in situ* of the oral cavity was the lack of a layered structural pattern. Diagnosis based on this criterion for dysplastic/malignant versus benign/reactive conditions achieved a sensitivity of 83% and specificity of 98% with an interobserver agreement value of 0.76. This study concluded that OCT, with high sensitivity and specificity combined with good interobserver agreement, is a promising imaging modality for non-invasive evaluation of tissue sites suspicious for high-grade dysplasia and cancer.

Other studies have utilized direct analysis of OCT scan profiles, rather than image-based criteria, as a means of delineating oral cancer lesions.[36,37] Numerical parameters from A-scan profiles were used as diagnostic criteria for OCT scans of oral lesions. The authors report that, consistently, the decay constant in the exponential fitting of the OCT signal intensity along depth decreased as the A-scan point moved laterally across the margin of a lesion. Additionally, the standard deviation of the SS-OCT signal intensity fluctuation in the A-scan increased significantly when the A-scan point was moved across the transition region between the normal and abnormal portions. The authors concluded that such parameters may well be useful for establishing an algorithm for detecting and mapping the margins of oral cancer lesions. Such a capability has huge clinical significance, because of the need to better define excisional margins during surgical removal of oral pre-malignancy and malignancy.

OCT is also under investigation for a host of innovative applications in related anatomical sites, such as the upper GI tract — especially Barrett's esophagus — and in Otolaryngology. These topics fall outside the scope of this paper, but we encourage readers to peruse some of these outstanding and informative papers.

Future Directions

Many research avenues are being explored in the quest to enhance and develop the imaging capabilities of OCT. Because the epithelium in the oral mucosa is so thin (200–500 μm), OCT imaging limitations in soft tissue of 2–3 mm are not an issue. Recent advances in OCT technology include 3-D topographic imaging capabilities, Fourier-domain OCT — permitting up to 100x faster acquisition of 3-D OCT images — and spectral OCT, providing enhanced image contrast and permitting spatially-resolved detection and quantification of changes within the tissues.[38–43] Polarization-sensitive OCT can be used to enhance collagen imaging,[44,45] and Doppler OCT provides a means for investigating vascularisation and altered tissue perfusion to identify vascular diagnostic benchmarks.[46,47] Advancements in OCT probe technologies, such as emerging MEMs-based techniques allow for the construction of miniaturized high-resolution probes that are suitable for endoscopic use[48–50] (Fig. 5). With several manufacturers working on developing OCT systems specifically targeted for use in the oral cavity, we anticipate that in the next five years several OCT devices optimized for use within the oral cavity will become available to clinicians.

With the rapid progress of improvements in OCT resolution capabilities there arises the need to enhance contrast levels in biological tissues. Various approaches are currently being explored. The use of gold nanoparticles as contrast agents to improve *in vivo* OCT images of oral dysplasia was explored in the standard hamster cheek pouch model for oral carcinogenesis.[51] Using a multimodal delivery system employing microneedles and ultrasound, gold nanoparticles were able to overcome biological barriers to achieve a 150% increase in OCT contrast levels (Fig. 7). Gold nanoparticles are biocompatible, easy to synthesize, functional with additional modalities and their optical resonance properties can be controlled over a broad range, making them a promising *in vivo* OCT contrast agent. The ability to use OCT contrast-enhancing agents clinically will certainly be an area of interest for future studies.

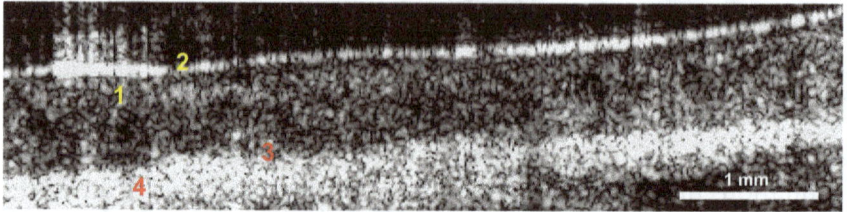

(a)

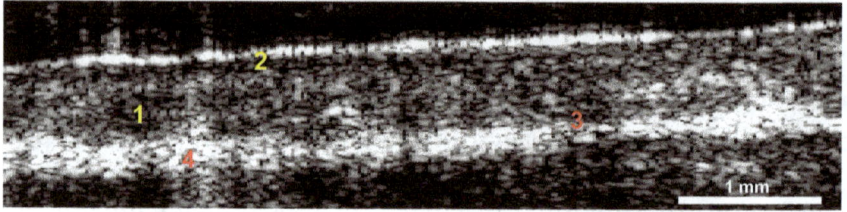

(b)

Fig. 7. Area of dysplasia in the hamster cheek pouch, imaged with OCT (a) before and (b) 50 minutes after gold nanoparticle application. Key: 1 — stratified squamous epithelium, 2 — keratinized epithelial surface layer, 3 — basement membrane, 4 — submucosa. In the second image, acquired 50 minutes after gold nanoparticles application, the microanatomy of the epithelium is more clearly defined and resolved, showing uneven keratinized surface layer, and downgrowth of pegs of the basement membrane into the underlying submucosa. The basement membrane is convoluted, but intact. Reproduced with permission from ChangSoo Kim, Yehchen Ahn, Young-Jik Kwan, Kenneth Lee, Zhongping Chen and Petra Wilder-Smith.

Conclusion

In 2006 it was estimated that more than 37,000 OCT scans were performed on a daily basis in the U.S., confirming the rapid translation of OCT technology from the laboratory to the patient.[52] It is always a challenge to translate technological advancements to clinical use. However, clinical trials have shown that OCT clearly has the potential to diagnose oral dysplasia and malignancy. The introduction of OCT imaging techniques to routine patient visits should be well-received by patients and clinicians alike, with the imaging protocol adding only a few minutes to visit duration. As stated by Tagg *et al.*, the hope is that simpler, less expensive and more quantitative procedures will provide the basis for earlier and more widespread testing of suspected lesions before invasive approaches are required.[53] Within the realm of oral medicine, the applications of OCT are extensive. They include the regular monitoring of potentially premalignant or malignant lesions, the detection of clinically nonapparent lesions, the entire scenario of detecting field cancerization and lesion recurrence, as well as mapping surgical margins. The key to improving patient survival and quality of life is very simple: very early and accurate detection. With oral cancer the 8th most common cancer among white males and the 6th most common cancer among black men in the United States, it is our hope that the incorporation of effective, non-invasive diagnostic modalities into oral health care will provide a direct, rapid, and cost effective means of better serving our patients.

References

1. The Oral Cancer Foundation. http://www.oralcancerfoundation.org.
2. American Cancer Society. (2009) *Cancer Facts and Figures 2009.* pp. 4; 16. American Cancer Society, Atlanta, Georgia.
3. Regezi JA, Sciubba JJ. (1993) *Oral Pathology.* WB Saunders Co., New York, pp. 77–90.
4. Lingen MW, Kalmar JR, Karrison T, Speight PM. (2008) Critical evaluation of diagnostic aids for the detection of oral cancer. *Oral Oncol* **44**: 10–22.
5. Rosin MP, Epstein JB, Berean K *et al.* (1997) The use of exfoliative cell samples to map clonal genetic alterations in the oral epithelium of high-risk patients. *Cancer Res* **57**: 5258–5260.
6. Jemal A, Siegel R, Ward E *et al.* (2008) Cancer statistics. *CA Cancer J Clin* **58**: 71–96.
7. Huang D, Swanson EA, Lin CP, Schuman JS, Stinson WG, Chang W, Hee MR, Flotte T, Gregory T, Puliafito C, Fujimoto JG. (1991) Optical coherence tomography. *Science* **254**: 1178–1181.
8. Fujimoto JG, Hee MR, Izatt JA, Boppart SA, Swanson EA, Lin CP *et al.* (1999) Biomedical imaging using optical coherent tomography. SPIE Vol. 3749, p. 402.

9. California Department of Health Services. Cancer Surveillance Section Annual Report, March 1999.

10. Slaughter DP, Southwick HW, Smejkal W. (1953) Field cancerization in oral stratified squamous epithelium. *Cancer* **6**: 963–968.

11. U.S. Department of Health and Human Services. A National Call to Action to Promote Oral Health http://www.nidr.nih.gov/sgr/nationalcalltoaction.htm. Rockville, MD, USA: Department of Health and Human Services, Public Health Service, Centers for Disease Control and Prevention, National Institutes of Health, National Institute of Dental and Craniofacial Research, May 2003. Report No.: 03–5303.

12. Sciubba JJ. (1999) Improving detection of precancerous and cancerous oral lesions: computer-assisted analysis of the oral brush biopsy. *J Am Dent Assoc* **130**: 1445–1457.

13. Epstein JB, Feldman R, Dolor RJ, Porter SR. (2003) The utility of tolonium chloride rinse in the diagnosis of recurrent or second primary cancers in patients with prior upper aerodigestive tract cancer. *Head Neck* **25**: 911–921.

14. Patton LL. (2003) The effectiveness of community-based visual screening and utility of adjunctive diagnostic aids in the early detection of oral cancer. *Oral Oncol* **39**: 708–723.

15. Onofre MA, Sposto MR, Navarro CM. (2001) Reliability of toluidine blue application in the detection of oral epithelial dysplasia and *in situ* and invasive squamous cell carcinomas. *Oral Surg Oral Med Oral Pathol Oral Radiol Endod* **91**: 535–540.

16. Farah CS, McCullough MJ. (2007) A pilot case control study on the efficacy of acetic acid wash and chemiluminescent illumination (ViziLite) in the visualization of oral mucosal white lesions. *Oral Oncol* **43**: 820–824.

17. Patton LP, Epstein JB and Kerr RA. (2008) Adjunctive techniques for oral cancer examination and lesion diagnosis: a systematic review of the literature. *J Am Dent Assoc* **139**: 896–905.

18. Epstein JB, Silverman S, Jr., Epstein JD, Lonky SA, Bride MA. (2008) Analysis of oral lesion biopsies identified and evaluated by visual examination, chemiluminescence and toluidine blue. *Oral Oncology* **44**: 538–544.

19. Poh CF, Ng SP, Williams PM, Zhang L, Laronde DM, Lane P, Macaulay C, Rosin MP. (2007) Direct fluorescence visualization of

clinically occult high-risk oral premalignant disease using a simple hand-held device. *Head Neck* **29**: 71–76.

20. Roblyer D, Kurachi C, Stepanek V, Williams MD, El-Naggar AK, Lee JJ, Gillenwater AM, Richards-Kortum R. (2009) Objective detection and delineation of oral neoplasia using autofluorescence imaging. *Cancer Prev Res* **2**: 423–431.

21. Choo-Smith LP, Edwards HG, Endtz HP *et al.* (2009) Medical applications of Raman spectroscopy: From proof of principle to clinical implementation. *Biopolymers* **67**: 1–9.

22. Bigio IJ, Mourant JR. (1997) Ultraviolet and visible spectroscopies for tissue diagnostics: Fluorescence spectroscopy and elastic-scattering spectroscopy. *Phys Med Biol* **42**: 803–814.

23. Sokolov K, Follen M, Richards-Kortum R. (2002) Optical spectroscopy for detection of neoplasia. *Curr Opin Chem Biol* **6**: 651–658.

24. Lane PM, Gilhuly T, Whitehead P *et al.* (2006) Simple device for the direct visualization of oral-cavity tissue fluorescence. *J Biomed Opt* **11**: 024006.

25. Schwarz RA, Gao W, Redden *et al.* (2009) Noninvasive evaluation of oral lesions using depth-sensitive optical spectroscopy simple device for the direct visualization of oral-cavity tissue fluorescence. *Cancer* **115**: 1669–1679.

26. De Veld DC, Witjes MJ, Sterenborg HJ, Roodenburg JL. (2009) The status of *in vivo* autofluorescence spectroscopy and imaging for oral oncology. *Oral Oncol* **41**: 117–131.

27. Rahman M, Chaturvedi P, Gillenwater AM, Richards-Kortum R. (2008) Low-cost, multimodal, portable screening system for early detection of oral cancer. *J Biomed Opt* **13**: 0305020.

28. White WM, Rajadhyaksha M, Gonzalez S, Fabian RL, and Anderson RR. (1999) Noninvasive imaging of human oral mucosa *in vivo* by confocal refectance microscopy. *Laryngoscope* **109**: 1709–1717.

29. Thong PS, Olivo M, Kho KW, Zheng W, Mancer K, Harris M, Soo KC. (2007) Laser confocal endomicroscopy as a novel technique for fluorescence diagnostic imaging of the oral cavity. *J Biomed Opt* **12**: 014007.

30. Maitland KC, Gillenwater AM, Williams MD, El-Naggar AK, Descour MR, Richards-Kortum RR. (2008) *In vivo* imaging of oral neoplasia using a miniaturized fiber optic confocal reflectance microscope. *Oral Oncol* **44**: 1059–1066.

31. Chang CJ, Wilder-Smith P. (2005) Topical application of photofrin for photodynamic diagnosis of oral neoplasms. *Plast Reconstr Surg* **115**: 1877–1886.

32. Leunig A, Mehlmann M, Betz C *et al.* (2001) Fluorescence staining of oral cancer using a topical application of 5-aminolevulinic acid: fluorescence microscopic studies. *J Photochem Photobiol B* **60**: 44–49.

33. Wilder-Smith P, Lee K, Guo S, Zhang J, Osann K, Chen Z *et al.* (2009) *In vivo* diagnosis of oral dysplasia and malignancy using optical coherence tomography: Preliminary studies in 50 patients. *Lasers Surg Med* **41**: 353–357.

34. Fomina JV, Gladkova ND, Snopova LB, Shakhova NM, Feldchtein FI, Myakov AV. (2004). *In vivo* OCT study of neoplastic alterations of the oral cavity mucosa. Proceedings SPIE, The International Society For Optical Engineering. **5316**: 41–47.

35. Ridgway JM, Armstrong WB, Guo S, Mahmood U, Su J, Jackson RP *et al.* (2006) *In Vivo* optical coherence tomography of the human oral cavity and oropharynx. *Arch Otolaryngol Head Neck Surg* **132**: 1074–1081.

36. Tsai MT, Lee HC, Lu CW, Wang YM, Lee CK, Yang CC *et al.* (2008) Delineation of an oral cancer lesion with swept-source optical coherence tomography. *J Biomed Optics* **13**: 044012.

37. Tsai MT, Lee HC, Lee CK, Yu CH, Chen HM, Chiang CP *et al.* (2008) Effective indicators for diagnosis of oral cancer using optical coherence tomography. *Opti Express* **16**: 15847–15862.

38. Drexler W, Fujimoto JG. (2008) *Optical Coherence Tomography Technology and Applications.* Springer, New York, USA.

39. Leitgeb RA, Hitzenberger CK, Fercher AF. (2003) Performance of fourier domain vs. time domain optical coherence tomography. *Opt Express* **11**: 889–894.

40. Yun S, Tearney G, de Boer J, Iftimia N, Bouma B. (2003) High-speed optical frequency-domain imaging. *Opt Express* **11**: 2953–2963.

41. Zhang J, Nelson JS, Chen Z. (2005) Removal of a mirror image and enhancement of the signal-to-noise ratio in Fourier-domain optical coherence tomography using an electro-optic phase modulator. *Opt Lett* **30**: 147–149.

42. Huber R, Wojtkowski M, Taira K, Fujimoto J, Hsu K. (2005) Amplified, frequency swept lasers for frequency domain reflectometry and OCT imaging: Design and scaling principles. *Opt Express* **13**: 3513–3528.

43. Kim JS, Ishikawa H, Gabriele ML, Wollstein G, Bilonick RA, Kagemann L, Fujimoto JG, Schuman JS. (2009) Retinal nerve fiber layer thickness measurement comparability between time domain optical coherence tomography (OCT) and spectral domain OCT. *Invest Ophthalmol Vis Sci* **51**: 896–902.

44. Lee SW, Yoo JY, Kang JH, Kang MS, Jung SH, Chong Y, Cha DS, Han KH, Kim BM. (2008) Optical diagnosis of cervical intraepithelial neoplasm (CIN) using polarization-sensitive optical coherence tomography. *Opt Express* **16**: 2709–2719.

45. Strasswimmer J, Pierce MC, Park BH, Neel V, de Boer JF. (2004) Polarization-sensitive optical coherence tomography of invasive basal cell carcinoma. *J Biomed Opt* **9**: 292–298.

46. Gonzalo N, Serruys PW, Piazza N, Regar E. (2009) Optical coherence tomography (OCT) in secondary revascularisation: Stent and graft assessment. *EuroIntervention* **5 Suppl D**: D93–D100.

47. Hanna NM, Waite W, Taylor K, Jung WG, Mukai D, Matheny E, Kreuter K, Wilder-Smith P, Brenner M, Chen Z. (2006) Feasibility of three-dimensional optical coherence tomography and optical Doppler tomography of malignancy in hamster cheek pouches. *Photomed Laser Surg* **24**: 402–409.

48. Ren H, Waltzer WC, Bhalla R, Liu J, Yuan Z, Lee CS, Darras F, Schulsinger D, Adler HL, Kim J, Mishail A, Pan Y. (2009) Diagnosis of bladder cancer with microelectromechanical systems-based cystoscopic optical coherence tomography. *Urology* **74**: 1351–1357.

49. Aguirre AD, Hertz PR, Chen Y, Fujimoto JG, Piyawattanametha W, Fan L, Wu MC. (2007) Two-axis MEMS scanning catheter for ultra-high resolution three-dimensional and en face imaging. *Opt Express* **15**: 2445–2453.

50. Su J, Zhang J, Yu L, Chen Z. (2007) *In vivo* three-dimensional microelectromechanical endoscopic swept source optical coherence tomography. *Opt Express* **15**(16).

51. Kim CS, Wilder-Smith P, Ahn YC, Liaw LH, Chen Z, Kwon YJ. (2009) Enhanced detection of early-stage oral cancer *in vivo* by optical coherence tomography using multimodal delivery of gold nanoparticles. *J Biomed Opt* **14**(3).
52. Walz M. (2006) Hot technologies for 2007. OCT: imaging of the future. *R&D Mag* 6(12).
53. Tagg R, Asadi-Zeydabadi M, Meyers AD. (2009) Biophotonic and other physical methods for characterizing oral mucosa. *Otolaryngologic Clinics of North America* **38**: 215–240.

Chapter 3

Optical Coherence Tomography in Laryngeal Cancer

Marcel Kraft* and Christoph Arens[†]

Introduction

Superficial lesions of the vocal folds and the laryngeal mucosa are generally first examined by indirect laryngoscopy, stroboscopy and voice analysis.[1] However, the following problems are encountered in the diagnostic investigation of laryngeal pathologies of unclear origin: On the one hand, indirect laryngoscopy and stroboscopy alone cannot always supply a specific diagnosis.[2] On the other hand, computed tomography and magnetic resonance imaging usually fail to provide a diagnosis in superficial lesions of the laryngeal mucosa.[3] Therefore, microlaryngoscopy with excisional biopsy under general anesthesia is customarily required for histological verification of diagnosis and to definitively rule out malignancy.[2]

Optical coherence tomography (OCT) is a new imaging technology, which measures near infrared light backscattered from within tissue. In a manner similar to ultrasound imaging, but at a much higher resolution, cross-sectional images are provided resembling vertical sections in histology.[4] Historically, OCT has first been used in other medical specialties such as ophthalmology and dermatology.

* Department of Otorhinolaryngology, Head and Neck Surgery, Kantonsspital AG, Aarau, Switzerland. E-mail: marcel.kraft@unibas.ch

[†] Department of Otorhinolaryngology, Head and Neck Surgery, University Hospital of Magdeburg, Magdeburg, Germany. E-mail: christoph.arens@med.ovgu.de

The development of fiber-based OCT devices has extended its application to the fields of gastroenterology and cardiology.[5] Recent investigations have shown that this noninvasive procedure can equally predict the exact extension, microstructure and possible malignancy of laryngeal lesions before surgery is performed. Therefore, this promising method opens new possibilities in the field of laryngology.[6]

The objective of this chapter is to summarize the current knowledge on OCT in the diagnostic investigation of laryngeal cancer and its precursor lesions.

Working Principle

Near infrared light generated by a superluminescent diode is split into a reference and a probe arm within a beam splitter (Fig. 1). The reference arm is reflected unchanged from a mirror, whereas the probe arm penetrates tissue where it is scattered by the different tissue layers. Both of them are superimposed again within the same beam splitter, giving rise to an interference signal. The latter can be discerned from a detector and evaluated at a computer, which allows a detailed insight into the structure of the investigated tissue.[4]

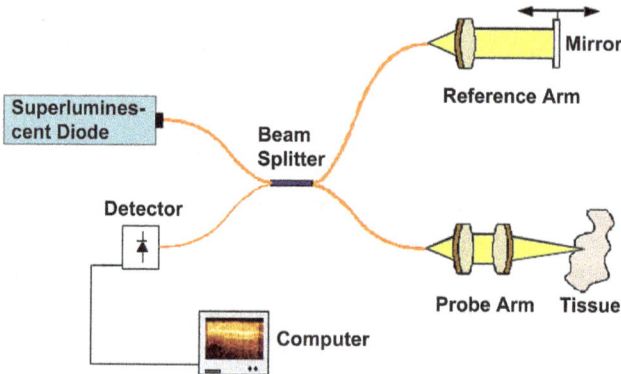

Fig. 1. Working principle of OCT. Near infrared light generated by a superluminescent diode is split into a reference and a probe arm within a beam splitter. After their reflection from a mirror or tissue respectively, both arms are superimposed again within the same beam splitter, giving rise to an interference signal. The latter can be discerned from a detector and evaluated at a computer, which allows a detailed insight into the structure of the investigated tissue.

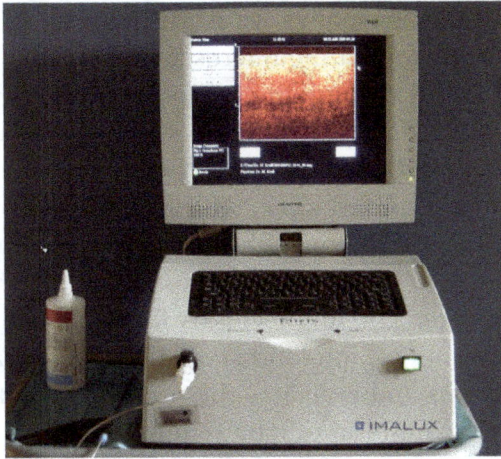

Fig. 2. Niris OCT imaging system. All equipment is placed on a single wheeled table, which can easily be moved to the operating theater on demand. The area under the probe is protected by a piece of foam rubber covered with a surgical drape to avoid damage to the sensitive OCT fiber.

In our institution, we currently use the Niris OCT imaging system (Imalux, Cleveland, OH) in contact mode for the diagnostic investigation of laryngeal pathologies of unclear origin (Fig. 2). This fiber optical interferometer is provided with a piezoelectric scanner for acquisition of cross-sectional images illustrated as a two-dimensional picture on an integrated monitor. From a technical point of view, a low-coherence light source with a central wavelength of 1300 nm is used yielding a spatial resolution of 10–20 μm and a depth of penetration between 1–2 mm depending on the turbidity of tissue. In such a manner, a 200 × 200 pixel image of 2 mm length is acquired within 1.5 seconds.[7] For OCT examination, the refractive index of the larynx should be set at 1.4 as proposed in the literature.[8]

Technique in the Larynx

Each patient with a laryngeal pathology of unclear origin should undergo indirect laryngoscopy, stroboscopy and voice analysis prior to surgery.[1] During microlaryngoscopy, the larynx is exposed using surgical laryngoscopes with suspension. The lesion is inspected under the operating microscope, palpated with the help of microforceps and documented photographically.[2] For further investigation, the Niris OCT imaging system is used before excisional biopsy or chordectomy is carried out. All equipment is placed on a single wheeled table, and the area under the probe is protected by a piece of foam rubber covered with a surgical drape (Fig. 2).

Generally, two persons are necessary for OCT monitoring of the larynx. The surgeon positions and keeps the probe steady, while the assistant documents and evaluates the procedure. With good communication, the acquisition of data requires less than 5 minutes of additional operating time. The OCT probe is introduced in a specially designed applicator (Karl Storz, Tuttlingen, Germany), which remarkably facilitates its positioning. The surgeon inserts the probe through the laryngoscope under microscopic control placing its tip in gentle contact with the region of interest. Systematic clockwise imaging of the larynx is performed beginning at the posterior third of the left vocal fold. Thereafter, the pathologic lesion is explored. Meanwhile, the assistant observes the monitor, starts and stops the scanning process with the use of a foot pedal, and precisely documents the recorded sites in a schematic drawing of the larynx. Finally, he letters and evaluates the OCT images and communicates his results, while the surgeon continues with the operation. On completion of the surgery, the probe is disinfected using Gigasept AF 4% solution (Schülke & Mayr, Norderstedt, Germany).[7]

OCT Findings

The normal vocal fold shows a thin translucent (in the OCT image brown) epithelium of approximately 50–150 μm thickness. This layer is followed by the increased scattering (in the OCT image yellow) lamina propria, which includes the Reinke's space with its embedded vessels and the vocal ligament. The vocal muscle is located much deeper and cannot be depicted due to the limited penetration depth of infrared light (Fig. 3).

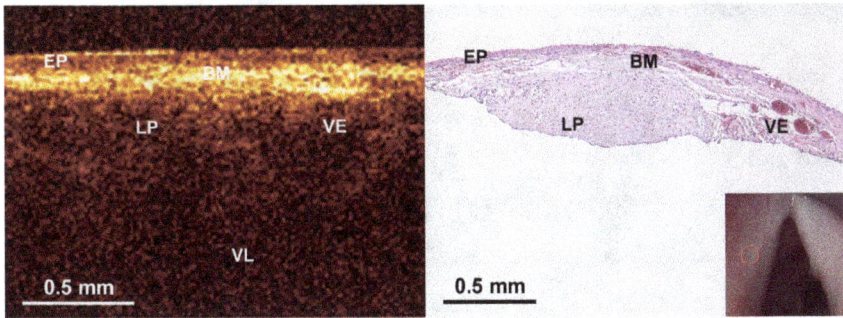

Fig. 3. Normal epithelium. The thin translucent epithelium (EP) with an intact basement membrane (BM) is followed by the increased scattering lamina propria (LP) with its embedded vessels (VE) and the vocal ligament (VL), which is hardly visible in the deep (HE stain, original magnification × 50).

Dysplasias demonstrate a thickened and increased scattering (in the OCT image yellow) epithelium. However, the basement membrane remains intact and can be recognized as a clear line in the deep. Superficial keratosis is identified by an intensely scattering (in the OCT image pale yellow) structure at the epithelial surface, whereas subepithelial inflammation is evidenced by translucent (in the OCT image dark brown) areas in the lamina propria just underneath the dysplastic epithelium (Fig. 4).

In invasive carcinomas, the basement membrane is no longer recognizable, and increased scattering (in the OCT image yellow) tumor plugs are seen reaching to deeper tissue layers. Subepithelial inflammation shows translucent (in the OCT image dark brown) areas in the lamina propria generally encompassing the tumor plugs. In our experience, penetration of the basement membrane is the most important OCT criterion for invasion. The appearance of vessels or inclusions within the increased scattering (in the OCT image yellow) tumor epithelium is another characteristic sign for malignancy. Unfortunately, the latter is not always present in cancers (Fig. 5).

With the exception of granulomas, benign lesions such as laryngeal cysts, polyps, Reinke's edemas, respiratory papillomas and scars offer no essential diagnostic difficulties because of their characteristic OCT image.[6]

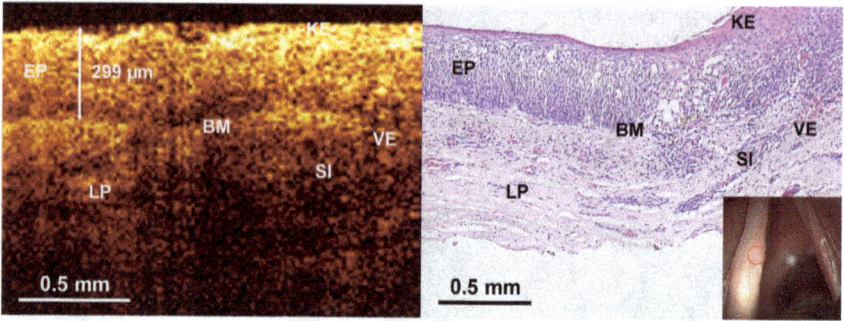

Fig. 4. Epithelial dysplasia. The thickened epithelium (EP) shows superficial keratosis (KE) and a clear basement membrane (BM) followed by subepithelial inflammation (SI) and dilated vessels (VE) in the lamina propria (LP). Morphometric measurement resulted in an epithelial thickness of 299 µm corresponding to dysplasia grade II. This was verified histologically after superficial chordectomy (HE stain, original magnification × 50).

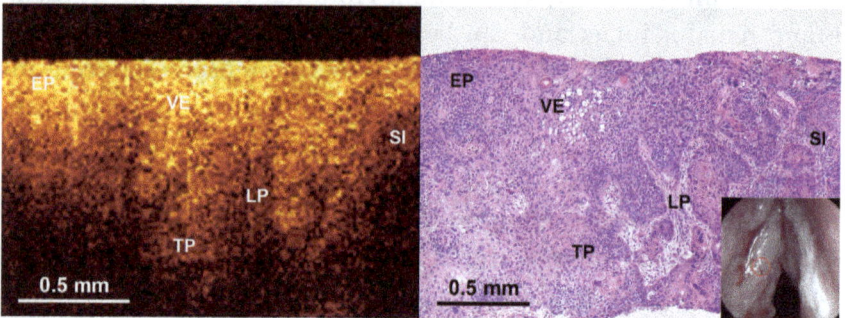

Fig. 5. Invasive carcinoma. The basement membrane is no longer recognizable, and tumor plugs (TP) are seen reaching to deeper tissue layers followed by subepithelial inflammation (SI) in the lamina propria (LP). The appearance of vessels (VE) within the tumor epithelium (EP) is another characteristic sign for malignancy. Our example demonstrates a T2-vocal fold carcinoma on the left (HE stain, original magnification × 50).

Grading of Dysplasia

Because of a spatial resolution of 10–20 µm, the Niris OCT imaging system is able to distinguish between different tissue layers, but cannot depict nuclear atypia as used in histopathological evaluation.

Therefore, other criteria have to be employed to determine the approximate grade of dysplasia.[6]

In previous investigations, we were able to demonstrate a progressive thickening from normal epithelium through the different grades of dysplasia to early invasive carcinoma.[9] In such a manner, moderate dysplasia shows a double increase, severe dysplasia and carcinoma *in situ* a triple increase, and early invasive carcinoma even a sixfold increase of the mean epithelial thickness compared to normal laryngeal mucosa. Benign lesions such as polyps, Reinke's edemas, chronic laryngitis and respiratory papillomas present with only a slight thickening of the epithelium, which is equally true for simple hyperplasia and mild dysplasia (Table 1).

Hence, with the help of OCT, the grade of dysplasia can be determined approximately by morphometric measurement of epithelial thickness (Fig. 4). A single exception is atrophic cancerous epithelium such as erythroplakia, which is related to pathologically dilated submucosal vessels and vascular atypias and not to epithelial thickening. As the latter is histologically often associated with carcinoma *in situ*, this could be a possible source of false negative results in OCT. However, this condition is easily

Table 1. Morphometric measurement of epithelial thickness ($n = 206$).

Laryngeal lesions	No. of cases	Mean value	Range
Normal epithelium	16	150 µm	110–186 µm
Benign lesions	41	240 µm	186–312 µm
Mild dysplasia	45	260 µm	208–317 µm
Moderate dysplasia	36	300 µm	261–353 µm
Severe dysplasia	26	400 µm	341–464 µm
Carcinoma in situ	18	500 µm	361–579 µm
Early invasive cancer	24	975 µm	805–1125 µm

Note: Measurements were performed on histological sections at 10 different sites perpendicular to the basement membrane capturing always the maximum width of the epithelium. Shrinkage of tissue samples due to paraffin fixation was compensated by adding 20% of each measured value.

visible during microlaryngoscopy. Therefore, in such lesions, the surgeon should proceed with excisional biopsy or chordectomy regardless of the OCT findings.[2]

Complications and Difficulties

To this time, no complications such as infection, bleeding, vocal fold damage or dental injury occurred during or after OCT imaging of the larynx.[10]

However, we encountered a few technical difficulties. For high-quality pictures without artifacts, the applicator should be propped against the inner surface of the laryngoscope. In lesions occurring at the free border of the vocal folds, a pair of microforceps placed underneath for stabilization facilitates the procedure. Alternatively, the free border of the vocal fold can be rotated upwards by gently pushing the microforceps into the laryngeal ventricle.

In addition, there are several difficulties in the interpretation of OCT images. Basically, a profound knowledge of vocal fold anatomy and histopathology is indispensable when performing this method of *in vivo* histology.[11] Distinct keratosis and acanthosis prevent deeper tissue layers from being evaluated further due to a strong light absorption. In such lesions, the basement membrane is not seen during OCT, and therefore subjacent cancer cannot be definitively excluded.[6] Subepithelial inflammation and circumscribed keratosis may occasionally lead to an overestimation of dysplasia. Hence, the presence of these changes should be recognized and subtracted from the total value of epithelial thickness to determine the correct grade of dysplasia.[10] Furthermore, ulcers and granulomas can be confused with invasive carcinomas due to the absence of a basement membrane. In these cases, a correct interpretation of OCT images is only possible through knowing the patient's clinical and microlaryngoscopic findings.[6] Finally, early invasive cancer cannot be safely delineated from severe dysplasia or carcinoma *in situ* due to the current spatial resolution of OCT. However, the therapeutic consequences in such lesions are generally the same.[7]

Research Results

In a prospective study, a total of 381 laryngeal lesions in 319 consecutive patients were examined during elective microlaryngoscopy. Clinical assessment under the operating microscope with and without OCT was compared with conventional histopathology after excisional biopsy.[10]

In such a manner, 116 malignant, 46 precancerous and 219 benign laryngeal lesions were classified according to the World Health Organization (WHO) to allow a comparison with the current literature (Table 2). In the evaluation of all pathologies, we found a significant benefit for OCT during microlaryngoscopy. In particular, the exact grade of dysplasia could be better determined with the help of OCT (Table 2). Additionally, OCT proved to be very helpful in identifying and excluding malignant tumors of the larynx (Table 3). Another value of OCT lies in monitoring the exact extension of laryngeal cancer and its precursor lesions allowing for a more appropriate surgical excision. Statistical significance was also demonstrated in predicting benign conditions, such as laryngeal cysts, polyps, Reinke's edemas, respiratory papillomas, granulomas and scars (Table 2).

Table 2. Comparison of specific and presumptive diagnosis ($n = 381$).

Laryngeal lesions	Specific diagnosis	Presumptive diagnosis (% of correct estimated cases by method)	
WHO-classification	Histopathology	MLS with OCT	MLS alone
Malignant lesions	116	91% ($p = 0.117$)	84%
Precanceroses	46	63%** ($p = 0.006$)	35%
Benign lesions	219	93%** ($p = 0.009$)	86%
Total lesions	381	89%** ($p < 0.001$)	79%

Abbreviations: WHO = World Health Organization; MLS = microlaryngoscopy; OCT = optical coherence tomography; * = statistical significance ($p < 0.05$); ** = high statistical significance ($p < 0.01$).

Note: Malignant lesions comprise invasive and early invasive cancer as well as carcinoma in situ, while precanceroses include simple hyperplasia and the different grades of dysplasia. All the other lesions were classified as benign conditions.

Table 3. Identification of malignant tumors ($n = 381$).

Method	Sensitivity	Specificity	Accuracy	Pos. predictive value	Neg. predictive value
MLS+OCT	97%*	95%	96%**	90%	99%*
	($p = 0.015$)	($p = 0.107$)	($p = 0.009$)	($p = 0.094$)	($p = 0.017$)
MLS alone	90%	92%	91%	83%	95%

Abbreviations: MLS = microlaryngoscopy; OCT = optical coherence tomography; * = statistical significance ($p < 0.05$); ** = high statistical significance ($p < 0.01$).

Note: Malignant lesions of the larynx are classified as positive, precanceroses and benign lesions as negative for the calculation of sensitivity, specificity, accuracy, positive and negative predictive value.

Hence, our data strongly suggest that OCT might be a useful adjunct to microlaryngoscopy for predicting the specific diagnosis in a variety of laryngeal lesions. In our opinion, this new technology has a great benefit for the inexperienced surgeon when dealing with precancerous and cancerous lesions, while the experienced laryngologist might profit less from OCT. In any case, this noninvasive procedure can help enormously in guiding biopsy and decision-making during laryngeal surgery.[6]

Comparison with the Literature

In the literature, there exist only a few publications on OCT imaging of the larynx.[12–27] Many of them are purely experimental studies with results that can only partly be applied to humans.[12–19] All the others are clinical trials with only a few histologically verified cases.[20–27] Reporting the findings of 26 patients with laryngeal cancer and its precursor lesions, Shakhov *et al.*[22] first demonstrated the potential value of OCT in laryngology. In a larger series of 82 patients, Wong *et al.*[23] performed OCT imaging in normal and benign laryngeal lesions. As only 73% of their patients underwent excisional biopsy, many interpretations are presumptive lacking histological confirmation.[23] In our opinion, a correct interpretation of OCT images is only possible through a comparison with conventional histopathology. In a limited study, Armstrong *et al.*[24] reviewed

OCT scans of 22 patients with laryngeal cancer and precursor lesions for the integrity of their basement membrane. Unfortunately, these authors did not indicate the percentage of correct preoperative diagnoses in their series nor did they mention if their evaluation was made in a blinded manner.[24] In our experience, the interpretation of OCT images is generally difficult and can strongly be influenced when the definitive histopathology is already known. Therefore, the real predictive value of OCT in laryngology can only be determined if an investigation is performed in a prospective and blinded manner. So far, our study is the only prospective clinical trial with a systematic examination of all major laryngeal pathologies, in which every lesion was confirmed by histopathology. For this reason, a real comparison of our results with the current literature is not as yet possible.[6]

Potential Applications

There are several advantages of using OCT in laryngology. First, this new technology can reliably predict the specific diagnosis in laryngeal lesions of unclear origin. Identifying and excluding malignancy during surgery is especially of great value, but OCT can also contribute to the oncologic aftercare. Second, OCT helps in determining tumor margins, which could one day result in a more appropriate surgical excision. Third, OCT allows for an optical analysis of tissue ("optical biopsy"). By no means are we suggesting that this promising method can presently replace conventional histopathology. Rather, OCT helps in guiding biopsy and has the potential of replacing frozen section for diagnostic purposes if performed in collaboration with a pathologist.[10]

An additional value of OCT lies in its combination with other diagnostic procedures.[6] Hence, microlaryngoscopy, autofluorescence and contact endoscopy enable us to evaluate the horizontal extension of a cancerous lesion, while high-frequency ultrasound, confocal laser microscopy and OCT basically allow for the vertical assessment of a tumor.[28] As monitoring the whole larynx in case of field cancerization is very time-consuming, autofluorescence endoscopy was found to be very helpful in detecting suspicious lesions followed by a specific OCT exploration.[7] In this context, endosonography helps in determining the exact extension of advanced carcinomas beyond 3 mm, while OCT especially qualifies for assessing laryngeal dysplasia and early invasive cancer up to 2 mm thickness due to its better resolution.[29] In our opinion, contact endoscopy is of little value as only the most superficial epithelial layers are reached. In contrast, confocal laser microscopy might become another important aid in the field of laryngology if the penetration depth can be technically increased.[30]

Conclusions

OCT is a simple and reliable aid in the early diagnosis of laryngeal disease, especially of laryngeal cancer and its precursor lesions. As OCT allows us to look a few millimeters in depth, it can help in guiding biopsy but presently not replace conventional histopathology. Currently, this is the only technique, which enables an intraoperative monitoring of any accessible mucosa within the larynx. By this means, OCT can contribute directly to therapy planning, as far as the evaluation of surgical margins and a more superficial excision of small cancers and precursor lesions are concerned. Additionally, OCT can easily be combined with other diagnostic procedures such as microlaryngoscopy and autofluorescence endoscopy. In such a manner, it holds the position of a complementary diagnostic tool in the investigation of laryngeal pathologies of unclear origin. The improved diagnostics as shown by our data justify the application of this noninvasive procedure in the field of laryngology. Besides, the Niris OCT imaging system used in our daily clinical practice is portable, less expensive than a common ultrasonograph and does not require technical support to run the system. Through further improvement of spatial resolution, penetration depth and image acquisition time, OCT might become an integral part in the investigation of laryngeal lesions in the near future. Recent developments of noncontact OCT devices integrated with an endoscope or surgical microscope will soon allow for an adequate office-based examination of the larynx. In such a manner, unnecessary laryngeal surgery or secondary operations could eventually be avoided in certain cases.

Acknowledgements

The authors would like to thank Warren Jacobs for correcting the manuscript and Imalux Corporation for supplying the Niris OCT imaging system.

Dedication

This chapter is dedicated to the 65th birthday of Professor Mihael Podvinec, who supported my research and enabled my education as an otolaryngologist.

References

1. Kaszuba SM, Garret CG. (2007) Strobovideolaryngoscopy and laboratory voice evaluation. *Otolaryngol Clin North Am* **40**: 991–1001.
2. Kleinsasser O. (1988) *Tumors of the larynx and hypopharynx.* Thieme Medical Publishers, New York: 124–150.
3. Becker M. (1998) Diagnosis and staging of laryngeal tumors with CT and MRI. *Radiologe* **38**: 93–100.
4. Huang D, Swanson EA, Lin CP, Schuman JS, Stinson WG, Chang W, Hee MR, Flotte T, Gregory K, Puliafito CA, Fujimoto JG. (1991) Optical coherence tomography. *Science* **254**: 1178–1181.
5. Bouma BE, Tearney GJ. (2002) Clinical imaging with optical coherence tomography. *Acad Radiol* **9**: 942–953.
6. Kraft M, Glanz H, von Gerlach S, Wisweh H, Lubatschowski H, Arens C. (2008) Clinical value of optical coherence tomography in laryngology. *Head Neck* **30**: 1628–1635.
7. Kraft M, Lüerssen K, Lubatschowski H, Glanz H, Arens C. (2007) Technique of optical coherence tomography of the larynx during microlaryngoscopy. *Laryngoscope* **117**: 950–952.

8. Tearney GJ, Brezinski ME, Southern JF, Bouma BE, Hee MR, Fujimoto JG. (1995) Determination of the refractive index of highly scattering human tissue by optical coherence tomography. *Opt Lett* **20**: 2258–2260.

9. Arens C, Glanz H, Wönckhaus J, Hersemeyer K, Kraft M. (2007) Histologic assessment of epithelial thickness in early laryngeal cancer or precursor lesions and its impact on endoscopic imaging. *Eur Arch Otorhinolaryngol* **264**: 645–649.

10. Kraft M, Glanz H, von Gerlach S, Wisweh H, Lubatschowski H, Arens C. (2010) Optical coherence tomography: Significance of a new method for assessing unclear laryngeal pathologies. *HNO* **58**: 472–479.

11. Hirano M, Sato K. (1993) *Histological Color Atlas of the Human Larynx.* Singular Publishing Group, San Diego: 1–58.

12. Bibas AG, Podoleanu AG, Cucu RG, Bonmarin M, Dobre GM, Ward VM, Odell E, Boxer A, Gleeson MJ, Jackson DA. (2004) 3-D optical coherence tomography of the laryngeal mucosa. *Clin Otolaryngol Allied Sci* **29**: 713–720.

13. Lüerssen K, Lubatschowski H, Gasse H, Koch R, Kuranov R, Ptok M. (2005) Characterization of vocal folds using optical coherence tomography. *Sprache Stimme Gehör* **29**: 35–39.

14. Burns JA, Zeitels SM, Anderson RR, Kobler JB, Pierce MC, de Boer JF. (2005) Imaging the mucosa of the human vocal fold with optical coherence tomography. *Ann Otol Rhinol Laryngol* **114**: 671–676.

15. Karamzadeh AM, Jackson R, Guo S, Ridgway JM, Wong HS, Ahuja GS, Chao MC, Liaw LL, Chen Z, Wong BJ. (2005) Characterization of submucosal lesions using optical coherence tomography in the rabbit subglottis. *Arch Otolaryngol Head Neck Surg* **131**: 499–504.

16. Nassif NA, Armstrong WB, de Boer JF, Wong BJ. (2005) Measurement of morphologic changes induced by trauma with the use of coherence tomography in porcine vocal cords. *Otolaryngol Head Neck Surg* **133**: 845–850.

17. Lüerssen K, Lubatschowski H, Radicke N, Ptok M. (2006) Optical characterization of vocal folds using optical coherence tomography. *Medical Laser Application* **21**: 185–190.

18. Lüerssen K, Lubatschowski H, Ursinus K, Gasse H, Koch R, Ptok M. (2006) Optical coherence tomography in the diagnosis of vocal folds. *HNO* **54**: 611–615.

19. Torkian BA, Guo S, Jahng AW, Liaw LH, Chen Z, Wong BJ. (2006) Noninvasive measurement of ablation crater size and thermal injury after CO_2 laser in the vocal cord with optical coherence tomography. *Otolaryngol Head Neck Surg* **134**: 86–91.

20. Sergeev AM, Gelikonov VM, Gelikonov GV, Feldchtein FI, Kuranov RV, Gladkova ND, Shakhova NM, Snopova LB, Shakhov AV, Kuznetzova IA, Denisenko AN, Pochinko VV, Chumakov YP, Streltzova OS. (1997) In vivo endoscopic OCT imaging of precancer and cancer states of human mucosa. *Opt Express* **1**: 432–440.

21. Feldchtein FI, Gelikonov GV, Gelikonov VM, Kuranov RV, Sergeev AM, Gladkova ND, Shakhov AV, Shakhova NM, Snopova LB, Terentjeva AB, Zagainova EV, Chumakov YP, Kuznetzova IA. (1998) Endoscopic applications of optical coherence tomography. *Opt Express* **3**: 257–270.

22. Shakhov AV, Terentjeva AB, Kamensky VA, Snopova LB, Gelikonov VM, Feldchtein FI, Sergeev AM. (2001) Optical coherence tomography monitoring for laser surgery of laryngeal carcinoma. *J Surg Oncol* **77**: 253–258.

23. Wong BJ, Jackson RP, Guo S, Ridgway JM, Mahmood U, Su J, Shibuya TY, Crumley RL, Gu M, Armstrong WB, Chen Z. (2005) In vivo optical coherence tomography of the human larynx: Normative and benign pathology in 82 patients. *Laryngoscope* **115**: 1904–1911.

24. Armstrong WB, Ridgway JM, Vokes DE, Guo S, Perez J, Jackson RP, Gu M, Crumley RL, Shibuya TY, Mahmood U, Chen Z, Wong BJ. (2006) Optical coherence tomography of laryngeal cancer. *Laryngoscope* **116**: 1107–1113.

25. Klein AM, Pierce MC, Zeitels SM, Anderson RR, Kobler JB, Shishkov M, de Boer JF. (2006) Imaging the human vocal folds in vivo with optical coherence tomography: A preliminary experience. *Ann Otol Rhinol Laryngol* **115**: 277–284.

26. Guo S, Hutchison R, Jackson RP, Kohli A, Sharp T, Orwin E, Haskell R, Chen Z, Wong BJ. (2006) Office-based optical coherence tomographic imaging of human vocal cords. *J Biomed Opt* **11**: 30501.

27. Vokes DE, Jackson R, Guo S, Perez JA, Su J, Ridgway JM, Armstrong WB, Chen Z, Wong BJ. (2008) Optical coherence tomography-enhanced microlaryngoscopy: Preliminary report of a noncontact optical

coherence tomography system integrated with a surgical microscope. *Ann Otol Rhinol Laryngol* **117**: 538–547.

28. Arens C, Glanz H, Dreyer T, Malzahn K. (2003) Compact endoscopy of the larynx. *Ann Otol Rhinol Laryngol* **112**: 113–119.

29. Arens C, Glanz H. (1999) Endoscopic high-frequency ultrasound of the larynx. *Eur Arch Otorhinolaryngol* **256**: 316–322.

30. Just T, Stave J, Boltze C, Wree A, Kramp B, Guthoff RF, Pau HW. (2006) Laser scanning microscopy of the human larynx mucosa: A preliminary, ex vivo study. *Laryngoscope* **116**: 1136–1141.

Chapter 4

Fluorescence Imaging of the Upper Aerodigestive Tract

Christian Stephan Betz[*,‡], *Andreas Leunig*[*,§]
and Christoph Arens[†,¶]

The Importance of Early Detection

Early detection of upper aerodigestive tract (UADT) cancer is of utmost importance for the prognosis of the patients concerned, as tumor stages show a negative correlation with survival. Unfortunately, early cancers usually remain silent for a rather long time, which often leads to a delayed first presentation to a doctor at a late tumor stage. This problem might possibly be solved by the introduction of extensive screening programs. However, even if those patients with premalignant or early malignant mucosal lesions do see a doctor or other trained personnel, the right diagnosis is not always correctly recognized due to the following reasons:

- Leukoplakias and erythroplasias (as typical precancerous lesions) are usually easily detectable by the naked eye, and it is generally

* Department of Otorhinolaryngology, Head & Neck Surgery, Grosshadern Medical Campus, Ludwig Maximilian University Munich, Marchioninistr. 15, 81377 Munich, Germany
† Department of Otorhinolaryngology, Head & Neck Surgery, Magdeburg University Hospitals, Otto von Guericke University Magdeburg, Leipziger Strasse 44, 39120 Magdeburg, Germany.
‡ E-mail: christian.betz@med.uni-muenchen.de
§ E-mail: andreas.leunig@med.uni-muenchen.de
¶ E-mail: christoph.arens@med.ovgu.de

recommended that upon detection one or more representative tissue biopsies are taken in order to establish a histopathological diagnosis. If no dysplastic or invasive growth is found, the lesion is either ablated using a CO_2 laser in defocused mode or just followed-up closely without any intervention. Otherwise (i.e. if a high grade dysplasia or invasive growth is present), the lesion is resected according to common guidelines in head and neck oncology with a safety margin of 0.5 cm. When applying this regimen, however, there remains the commonly underestimated risk of false negative results by mis-sampling — i.e. missing (pre)malignantly transformed areas within the often widespread lesions, which can occur side by side to regularly layered but only hyperkeratotic mucosa and have the same macroscopic appearance.

- Early SCCs that have not developed from obvious precancerous conditions, on the other hand, may either appear as an ulcer, erosion or even only slightly roughened healthy mucosa and do not yet display typical morphologic features of malignant tumours. Therefore, it may easily be mistaken for some chronic inflammatory condition or not be noticed at all.

As stated above, a delay of diagnosis is, however, indirectly correlated with the prognosis of the patients concerned. Therefore, it seems highly important to improve on early detection as well as identification and demarcation of these (pre)malignant lesions. In order to pursue this goal, fluorescence imaging has been widely tested and applied during the last two decades.

Biophysical Basics

Autofluorescence imaging

Autofluorescence imaging is based on a two-dimensional imaging of fluorescence emitted to the surface by the main tissue fluorophores following adequate excitation. The spectral characteristics of the endogenous fluorophores that chiefly determine the autofluorescent properties of tissues are listed in Table 1.

Table 1. Excitation and emission peaks of the most important endogenous fluorophores (adapted from Refs. 1 and 2).

Fluorophore	Cytological occurrence	Histological occurrence	Excitation peaks	Emission peaks
NADH	Intracellular	all layers	260 + 350 nm	440 + 450 nm
FAD	Intracellular	all layers	450 nm	515 nm
collagen	Extracellular	submucosa	330 nm	390 nm
elastin	Extracellular	submucosa	350 nm	420 nm
keratin	Intra-/extracellular	mucosa	340 nm	430 nm
porphyrins	Intracellular	superficial	405 nm	635 nm

When excited by an appropriate light stimulus, most of these compounds emit visible fluorescent light in the violet to green region of the spectrum. Additionally, there are intermediate products of haem biosynthesis (porphyrins) that produce a red fluorescent emission. The aromatic amino acid tryptophan has an excitation band in the ultraviolet range; as it is not excited in those autofluorescence imaging (AFI) systems that are commercially available, it is not listed in the table.

While the porphyrins display a very narrow range of excitation and emission properties, FAD and NADH as well as the fluorescent structural proteins collagen, elastin, and keratin have relatively broad excitation and emission bands. Thus, all of these molecules (except tryptophan) are excitable to some degree by exposure to short-wave visible light, which is used in all of the commercially available AFI systems.

Figure 1 shows a greatly simplified schematic diagram of the complex principles that underlie the autofluorescence diagnosis of malignant tumors. Over all, the green autofluorescence seems to be regularly diminished in cancerous tissue if compared to regular mucosa.

There are thought to be mainly three causes for this:

1. Neoplastic tissue in the UADT contains smaller concentrations of the intracellular fluorophore enzymes NADH and FAD than

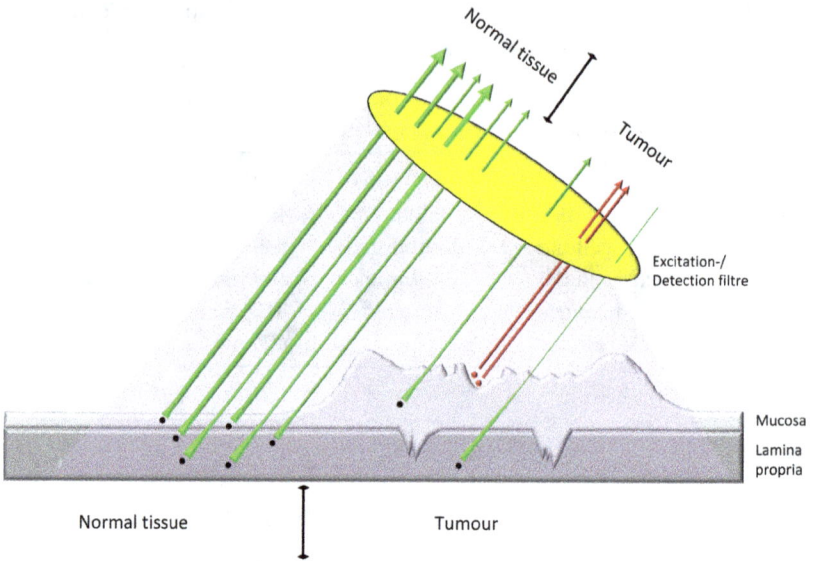

Fig. 1. Schematic representation of AFI based on the excitation of endogenous fluorophores in different tissue layers. Tumor tissue (right) differs from normal tissue in the following respects: (1) It contains smaller amounts of green-fluorescent fluorophores. (2) The intensity of the fluorescence is decreased by the thickened epithelial layer. (3) There is a variable amount of porphyrin-containing bacterial deposits that produce a red fluorescence.

normal tissue. This could account for the lower intensity of cellular autofluorescence found in tumor tissue compared with normal tissue. Additionally, tumor tissue (at least in the UADT) appears to have greater mitochondrial activity, and thus greater aerobic metabolic activity, than normal tissue. This shifts the balance from fluorescing NADH and FAD to the non-fluorescent compounds NAD and FADH.

2. The varying distribution of endogenous fluorophores in different tissue layers (Table 1) has an important bearing on fluorescence. This is because (pre)malignant thickening of the epithelial layer can make it more difficult for the excitation light to enter and fluorescent signals to leave the submucous tissue layers. This

minimizes the contribution of the structural proteins collagen and elastin to the total fluorescent emission.

3. A third factor that might also attribute to a reduction of auto-fluorescence signals over neoplastic mucosal lesions to a certain extent is an increased absorption of the excitation light by hemoglobin. This phenomenon is so far mostly acknowledged for AFI in the field of pulmonology.

Thus, with autofluorescence excitation in the near-UV and visible short-wave spectral range, all three of these phenomena tend to weaken the green fluorescent emissions that are measured in tumor tissue as compared with normal tissue (i.e. malignant lesions are demarcated by a darker shade of green). The structural protein keratin emits a pale green to whitish fluorescence in response to photoexcitation. Unfortunately, this protein cannot contribute to autofluorescence diagnosis because it does not seem to have a distinct tumor specificity. Porphyrins emit a red fluorescence, but they are of questionable value in the diagnosis of upper aerodigestive neoplasms. The principal source of this fluorescence is porphyrin-producing bacterial strains that may colonize the ulcerating surface of tumors. But these organisms show a somewhat inconsistent and nonhomogeneous distribution on mucosal neoplasms in the upper aerodigestive tract. Moreover, other tumor-nonspecific bacterial deposits, like those occurring at gingival margins or on the surface of the tongue, frequently produce a confusing, false-positive red fluorescence.

Enhanced fluorescence imaging (EFI)

If additional, tumor-specific fluorophores or their precursors are exogenously applied, then the optical technique is termed enhanced fluorescence imaging (EFI). 5-aminolevulinic acid (5-ALA) as a natural precursor of fluorescent porphyrins and haem is thereby most commonly employed. The formation of 5-ALA (and thus the biosynthesis of haem) is usually regulated by a negative feedback mechanism, which is surpassed if it is given directly. Following topical application (e.g. as a mouthwash for examinations of the oral cavity), it is preferentially taken up by (pre)malignant lesions and converted into strongly red fluorescent Protoporphyrin IX (PPIX). The most probable reasons therefore are: 1) enhanced penetration of 5-ALA into neoplastic tissue due to an impaired superficial lipid barrier, 2) improved conversion of 5-ALA into PPIX in malignant cells due to increased enzyme-activities and 3) retention of PPIX in malignant cells due to a reduced conversion into haem.

The fluorescence imaging following topical application of 5-ALA and generally using the same technical equipment as for AFI differs from that in a way illustrated in Fig. 2. Cancerous tissue is thereby not demarcated by a darker shade of green, but a bright red. This has been termed a "streetlight contrast" before (red: cancer, green: normal tissue).

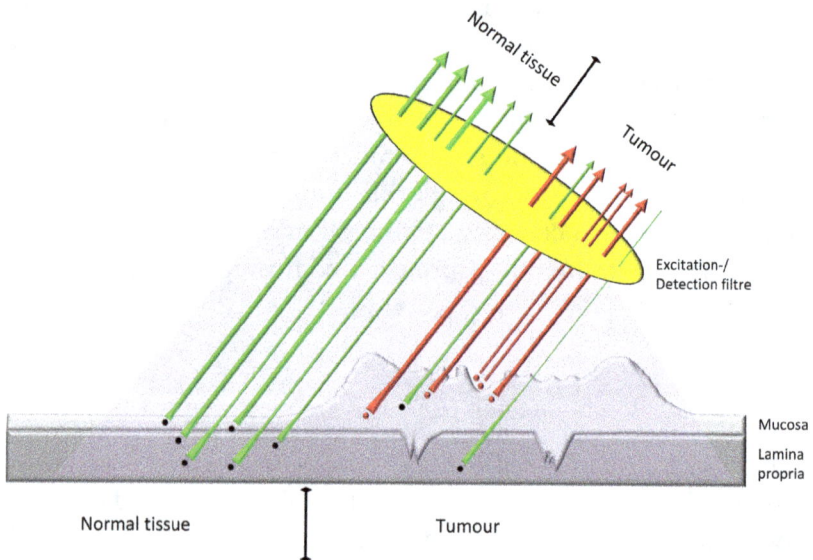

Fig. 2. Schematic representation of EFI using 5-ALA induced PPIX. In addition to the principles described and illustrated for AFI in Fig. 5, which are also true for EFI, the right sided cancerous tissue contains a much red-fluorescing PPIX than the left sided healthy tissue.

Equipment and Examination

Currently, there are three commercially available systems on the market for clinical fluorescence imaging in the head and neck field: the D-Light C/AF System (KARL STORZ GmbH, Tuttlingen, Germany), the DAFE System (Richard Wolf, Knittlingen, Germany) and the VELscope (LED Dental, White Rock, British Columbia, Canada). The first two of those are endoscopic systems that are (besides pulmonology) specifically marketed for use in the UADT; the third is a smaller, hand held device that is mainly advertised for oral examinations in dental offices. All 3 systems provide white-light illumination and fluorescence excitation in the blue/near-UV.

The D-Light C/AF System is the only one suitable for AFI as well as EFI — the other two systems are for AFI only. Whereas the systems by STORZ and the LED Dental perform a direct visualization of the detected fluorescence signals on a screen, the output of the Wolf system is a superimposed display of two separate detection ranges (500–590 nm and 600–700 nm) in false colors.

The main differences of the three systems are presented in Table 2.

Both methods can be performed under general anaesthesia as well as in awake patients. For AFI, no special preparation is necessary,

Table 2. Comparison of commercially available systems for fluorescence imaging of the UADT.

System	Suitable for	Excitation	Detection	Marketed for	Display
D-Light C/AF system	AFI + EFI	375–440 nm	>475 nm	UADT + lung	Direct display of fluorescence
DAFE system	AFI	390–460 nm	500–590 + 600–700 nm	UADT + lung	False color image
VELscope	AFI	400–460 nm	>475 nm	Oral cavity	Direct display of fluorescence

and the systems can easily be switched from white light into fluorescence mode.

EFI is performed after application and incubation of 5-ALA. For oral and oropharyngeal examinations, the substance is applied as a rinsing solution (200 mg 5-ALA dissolved in 50 ml of H_2O) for 15 minutes. For laryngeal sites, an inhalation of 5-ALA (30 mg 5-ALA dissolved in 5 ml normal saline) is performed. Following application, an incubation period of 1–2 hours should be kept before the examination to achieve the best possible results.

Fluorescence Imaging — General Statements

Autofluorescence imaging (AFI)

Autofluorescence findings of the UADT can generally be categorized into three distinct groups:

- Normal autofluorescence: Normally stratified, nonkeratinized squamous epithelium with a normal submucosa shows a typical homogeneous, pale green fluorescence (Fig. 3). Superficial capillary blood vessels show greater contrast with surrounding green fluorescence than in normal endoscopy (white light imaging = WLI) due to light absorption by hemoglobin.

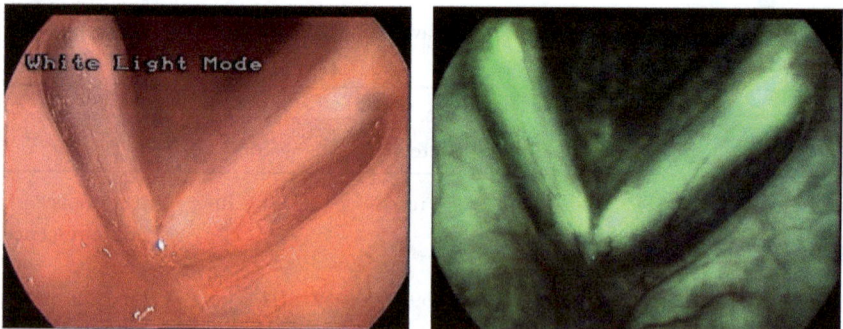

Fig. 3. Example for a finding with normal tissue autofluorescence from a healthy inner larynx on the right and corresponding white light image on the left.

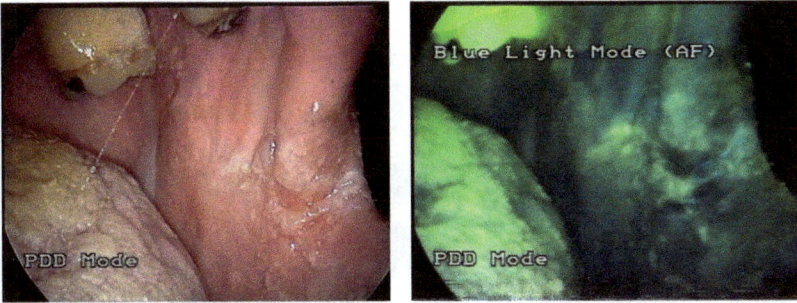

Fig. 4. Hyperkeratotic, highly dysplastic lesion of the left oral commissure (left: WLI, right: AFI).

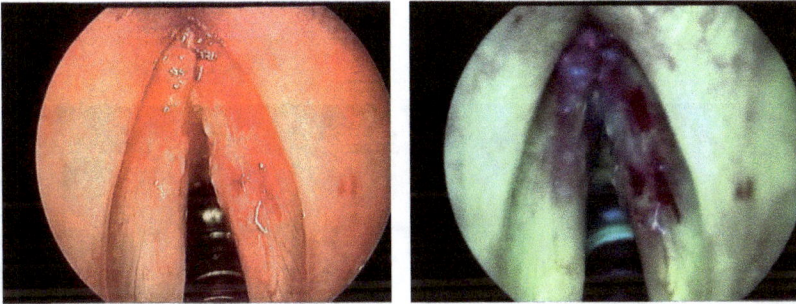

Fig. 5. Direct laryngoscopic appearance of a T1b glottic carcinoma of the anterior commissure. In AFI, the tumor contrasts sharply with its normal surroundings (left: WLI; right: AFI).

- Increased autofluorescence: In the case of increased superficial keratinization, the strong fluorescence of keratin becomes increasingly apparent. These areas show an intense green to whitish fluorescence when examined using AFI. A similar appearance is found in superficially heavily keratinizing (pre)malignant mucosal lesions (Fig. 4), "masking" the normal autofluorescence appearance of (pre)cancer.
- Decreased autofluorescence: (Pre)malignant mucosal lesions as well as benign lesions that are associated with a thickening of the mucosal layer are usually highlighted by sharply circumscribed areas of decreased autofluorescence (Figs. 5 and 6)

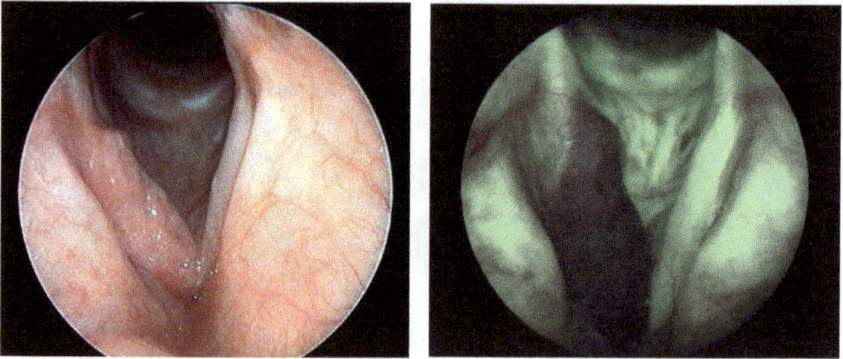

Fig. 6. Glottic laryngeal carcinoma (pT1a) visualized by indirect laryngoscopy. During AFI (right), the tumor shows a better contrast to the surrounding tissue than during WLI (left).

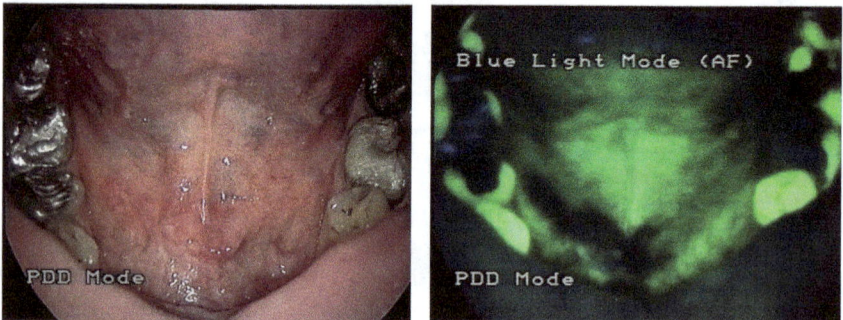

Fig. 7. Example of a premalignant lesion (CIS of the right floor of the mouth) in WLI on the left and AFI on the right side.

that sometimes contained reddish spots. The tumor grading does not influence the AF appearance, and tumor precursors (e.g. D III on Fig. 7) cannot be distinguished from invasive cancers based on their autofluorescent properties. Inflammatory thickenings of the mucosa and other changes such as benign neoplasms, submucous hemorrhages and hyperemia and fibrotic changes due to previous surgery also causes decreased

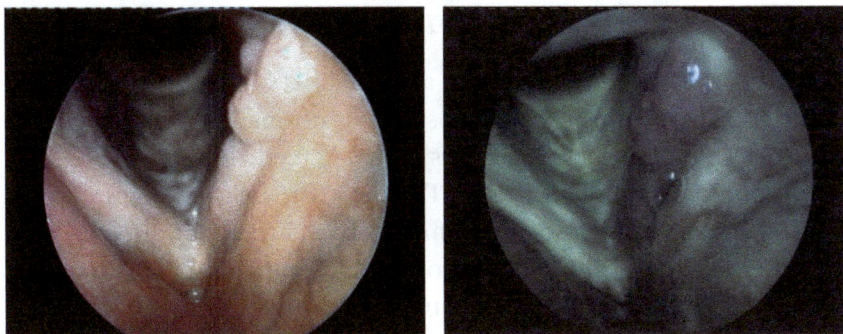

Fig. 8. HPV-associated papillomatosis of the left vocal fold visualized by indirect laryngoscopy (left: WLI; right: AFI). The papilloma shows a reduction in autofluorescence intensity (false positive).

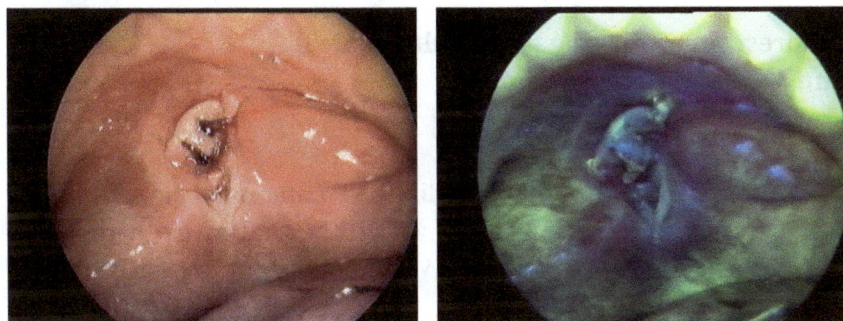

Fig. 9. Example for a "false positive" reduction in autofluorescence intensities on the right side and corresponding WLI on the left side in a patient with incipient fibrotic scar formation and local inflammatory reaction seven days after the resection of a pT1 G3 squamous cell carcinoma of the anterior floor of the mouth.

autofluorescence (Figs. 8 and 9). However, inflammatory and other non-neoplastic lesions usually cause less well-defined areas of decreased autofluorescence than neoplastic changes, so that a differentiation is often possible for the experienced investigator.

Enhanced fluorescence imaging (EFI)

Following adequate application and incubation of 5-ALA, normal tissue usually remains green (i.e. showing normal autofluorescence), whereas neoplastic tissue displays a strong, red fluorescence that is based on an accumulation of PPIX (Fig. 10). The red-to-green contrast is usually easily visible, similar to a streetlight. As for AFI, different gradings or stages cannot be differentiated by this method. As well, inflammatory or other benign mucosal lesions tend to induce false positive results, as do areas with a high grade of bacterial colonization (e.g. gingival margins, dorsum of tongue, etc.; Fig. 11). Pre-irradiated tissue shows a generalized, unspecific red fluorescence following application of 5-ALA, so EFI doesn't seem recommendable for these cases.

Fluorescence Imaging — Results in Oral and Oropharyngeal Lesions

Autofluorescence imaging (AFI)

In our own, ongoing study including 128 patients so far with highly suspicious or cancerous changes within the oral cavity or oropharynx, the additional use of AFI (to WLI) enabled us to clearly identify

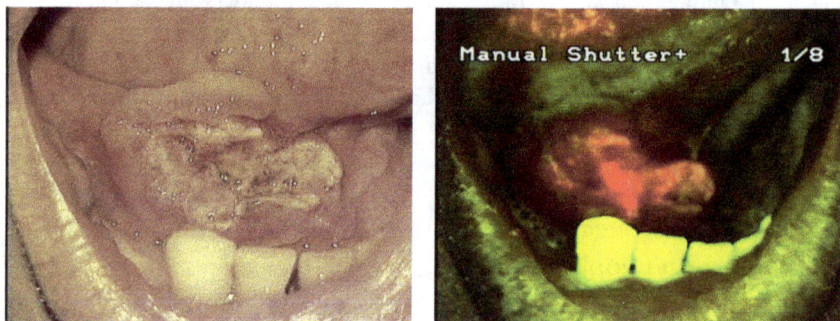

Fig. 10. Typical "streetlight contrast" (red: tumor; green: normal tissue) when using EFI at the example of a T2 squamous cell carcinoma of the floor of the mouth (left: WLI; right: EFI).

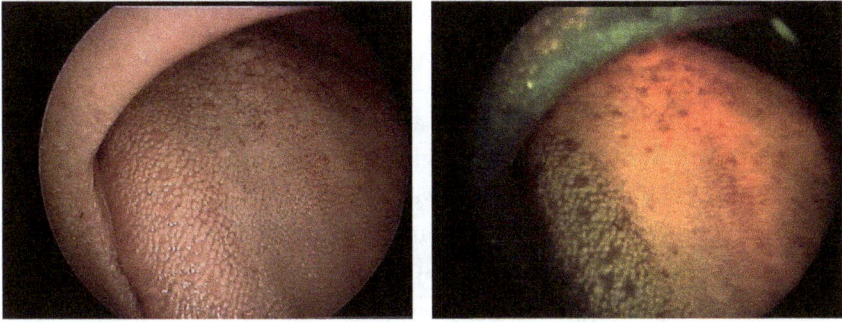

Fig. 11. False positive red fluorescence as a typical finding at dorsums of tongue (left: WLI; right: EFI).

a surplus of ten (pre)malignant mucosal lesions. The (pre)malignant lesions were thereby subjectively clearly identifiable in 89.5% (WLI) vs. 84.2% (AFI) and superficially delimitable in 37.9% (WLI) vs. 57.9% (AFI). Histopathological correlation of tissue biopsies from suspicious areas ($n = 236$) yielded a sensitivity of 95.8% (WLI) vs. 93.3% (AFI). Due to the fact that no random tissue biopsies were taken, the true selectivity couldn't be correctly determined even though it felt rather low.

Enhanced fluorescence imaging (EFI)

In previously published data from our group[3] on 68 patients with either high suspicion for or histologically proven SCC of the oral cavity or oropharynx, an additional eight (pre)malignant lesions were clearly identified only because EFI was used in conjunction with WLI. The rate of lesions that were subjectively clearly identifiable was 89.7% for both methods, whereas the rate of a good superficial demarcation of the lesions was 63.2% for EFI and 35.3% for WLI. The sensitivity for the histopathological correlation for $n = 199$ tissue biopsies from suspicious areas was determined as 99.2% for WLI and 100% for EFI. Again, reproducible numbers concerning the specificity of the method couldn't be calculated for the same reasons mentioned above.

Fluorescence Imaging — Results in Laryngeal Lesions

Autofluorescence imaging (AFI)

In 111 patients we compared AFI to WLI during direct and indirect laryngoscopy. With histopathologic diagnosis as standard, diagnostic accuracy by indirect WLI was 78%, that of AFI 90%. The sensitivity by indirect WLI for detection of precancerous as well as cancerous lesions was 85%, and specificity was 69%. The sensitivity by indirect AFI amounted to 91% and specificity to 87%. Indirect AFI improved the sensitivity by 6% and the specificity by 18% compared to normal WLI.

On the other hand, the diagnostic accuracy by WLI during microlaryngoscopy was 89%, that of AFI 95%. With histopathology as control, direct WLI of the larynx achieved a sensitivity of 92% and a specificity of 86%. The sensitivity of direct AFI laryngoscopy was 97% and specificity 92%. Diagnostic accuracy improved from indirect to direct WLI by 11% and AFI by 6%.

Secondary to scarring and chronic inflammation, false positive cases could be observed during direct and indirect laryngoscopy. Furthermore, false negative cases can be caused by a keratosis shielding the cancerous lesion in the basal cell layers (umbrella-effect).

Enhanced fluorescence imaging (EFI)

In a previously published study,[4] 56 patients with suspicious, flat epithelial lesions ($n = 7$ epithelial hyperplasias, $n = 15$ mild epithelial dysplasias, $n = 7$ moderate epithelial dysplasias, $n = 10$ severe epithelial dysplasia, $n = 17$ invasive carcinomas) were prospectively investigated using EFI following application of 5-ALA via inhalation two hours prior. Using the combined information obtained from WLI and EFI, a correct diagnosis could be obtained in 84% of cases, and the sensitivity was 97% with a corresponding specificity of 64%. As well, EFI was able to identify all invasive cancers.

Discussion

General points

Whereas the advantages of screening for early UADT cancer especially in high-risk populations seems beyond all question, the methods by which this is best achieved are not clear yet. The best and most commonly tested technique with a proven benefit so far is regular oral screening by trained health workers.[5] Other methods such as toluidine blue staining or fluorescence imaging, however, seem to be advantageous at least in theory. Yet, a Cochrane Review by Kujan *et al.* from 2006 found "no robust evidence to suggest that these are either beneficial or harmful".[6] The authors conclude that "future high quality studies to assess the efficacy, effectiveness and costs of screening are required for the best use of public health resources".

Three recent reviews have looked at the different techniques (including AFI) that are currently tested and applied as an adjunct for UADT cancer diagnosis.[7–9] The common feeling is that — as RCTs concerning their effectiveness are still missing for all of them — all these methods seem promising in theory but that the "tantalizing implication that such technologies may improve detection of oral cancers and precancers beyond conventional oral examination alone has yet to be rigorously confirmed".[8] Nevertheless, it is a commonly expressed feeling that fluorescence imaging is a useful, easy to perform and harmless diagnostic adjunct for UADT cancer screening.

Comparison of fluorescence Imaging to other novel screening techniques

Direct comparisons of toluidine blue or acetowhite staining (or the combination of both) with AFI or EFI have not been performed so far. Therefore, any comparative statements may be of limited value. If the extensively published data on toluidine blue staining is compared to our findings (reviewed in Ref. 10), however, we feel that fluorescence imaging might be slightly more sensitive to detect mucosal lesions especially in cases of dysplasia. Nevertheless, it seems worth undertaking a comparative study in order to finally solve this question.

Unlike toluidine blue and acetowhite staining, contact endoscopy following application of methylene blue has been used in conjunction with AFI in one of our larger *in vivo* studies.[11] In 73 out of 83 cases (88%) with dysplastic or malignant laryngeal lesions, the results from a combination of both techniques (which we termed "compact endoscopy") corresponded to those of histopathology (including the exact degree of dysplasia). It could be concluded from the study that both methods complement each other to a fair degree, and advocate compact endoscopy for a more accurate assessment of laryngeal cancer and its preceding lesions during microlaryngoscopy. Even though these are highly interesting results and other authors have reported promising results when applying contact endoscopy alone to UADT lesions, the usefulness of this method in conjunction with AFI (or EFI) is so far based on a single study only and would need to be verified by other *in vivo* trials.

Apart from a few case reports, there are only two studies on Narrow Band Imaging (NBI) published in medical literature so far. In one study by Watanabe *et al.*, the UADTs of 217 patients with oesophageal cancer were screened using NBI in combination with WLI.[12] They found 6 second primaries, whereas the authors claim that four of those were only detected due to the use of NBI. In the other trial published by Ugumori *et al.*,[13] 50 early SCCs of the UADT in 29 patients were investigated using WLI and NBI under local anaesthesia. The lesions were thereby subjectively better identifiable and delimitable against normal mucosa than with WLI alone. As the data on the usage of NBI in the UADT is very limited so far, the true value of the method can neither yet be fully assessed nor can it easily be compared to fluorescence imaging at the moment.

Comparison of different fluorescence imaging methods and systems

Both fluorescence imaging methods have been reported to yield a highly sensitivity concerning the identification and delimitation of UADT lesions with values ranging between 83 and 99%.[3,4,14–19] Yet, these methods seem to be generally lacking specificity. In a comparative study of AFI and EFI using a first generation AF system in 2002, our group found a slight advantage of EFI over AFI concerning identification of both tumor and borders.[3] When changing over to a newer system generation afterwards, however, we've seen a much better performance of AFI, which now seemed subjectively equivalent to EFI. However, we have not yet endeavoured a direct comparison of AFI and EFI with the second generation system, but this is planned for future research.

The publically available data on the use of the DAFE system (Richard Wolf GmbH, Knittlingen, Germany) in the head and neck region is limited to two publications so far (with one of them being in Polish).[20,21] Yet, and even though it processes the detected fluorescence before displaying it in false colors unlike the STORZ system, the reported results from a study on 96 early (T1 and T2) carcinomas of the UADT were very similar to the ones found in the current investigation. The authors state that, while all known lesions showed autofluorescence abnormalities, they were able to detect a surplus of nine previously undetected (pre)malignant lesions and found an otherwise invisible, superficial spread in two lesions. Even though a direct comparison of the two systems seems mandatory in the long run, the two systems seem to perform equally for all that is known to date.

Before the company (or its intellectual property, respectively) was taken over by Novadaq Technologies (Ontario, Canada) in early 2007, the so-called LIFE System by the former Xillix Technologies Corporation (Richmond, Canada) was also assessed for UADT tumor diagnosis, even though it was originally marketed for pulmonary and gastrointestinal use only.[22] Similar to the DAFE System, it also displayed a processed version of the obtained autofluorescence signals rather than the original fluorescence findings, but in this case tissue excitation was achieved with a helium-cadmium laser at 442 nm. In their small study, Kulapaditharom and Boonkitticharoen reported that they were able to identify all 16 cancerous lesions (including two occult cancers) in a group of 25 patients with suspicious lesions, while white light endoscopy only achieved an 87.5% detection rate. The results again seem similar to those achieved with the STORZ System, yet a direct comparison has not been performed yet and will supposedly also not be in the future as the system has been taken off the market.

Similar to the DAFE system, publications about the VELscope (LED Dental, White Rock, British Columbia, Canada) are scarce.[23–25] In one initial study, 50 (pre)malignant oral lesions from 44 patients were investigated using the VELscope, and the authors report a sensitivity of 98% and a specificity of 100% as for their correct detection.[23] As the system is primarily marketed for a general dental market, another one of the studies[25] was focused on this clientele. In two consecutive years, an oral cancer screening was performed on a random fraction of patients (1st year: 959/2133 patients, 2nd year 905/2029 patients) visiting a general dental practice; in the first year using WLI alone, in the second year with WLI and VELscope AFI. The detected prevalence of mucosal lesions was 0.83% ($n = 8$) or 1.30% ($n = 12$), respectively, whereas none of those found in the first year and 8 (out of 12) found in the second year were finally diagnosed as epithelial dysplasia. The authors conclude that a "routine incorporation of the VELscope in the examination protocol for low-risk adolescents and adults in a general dental practice" seems "useful in identifying occult, potentially premalignant lesions". In the third paper published by Poh *et al.*,[24] the VELscope was used on 20 oral cancer cases in addition to WLI, and biopsies from tumour centres ($n = 20$) and borders ($n = 102$) were correlated to fluorescence findings. In 19 out of 20 cases, the tumour borders using AFI extended beyond the clinically visible tumour. The sensitivity of the biopsies from the tumour boundaries was 97.3%, the specificity was 94.2%. The authors conclude that AFI appears able to "identify subclinical high-risk fields with cancerous and precancerous changes". Again, these results seem similar to those obtainable for AFI when using the STORZ system, yet a direct comparison seems indicated to find out about possible subtle differences.

Conclusion

In summary, AFI and EFI seem to be helpful, sensitive, yet rather unspecific adjunctive tools for the detection and superficial delimitation of (pre)malignant lesions of the UADT. In comparison, both seem to perform at least as good as other adjunct methods for UADT cancer screening. Both AFI and EFI are easy to perform, non-invasive and well accepted by the patients. Yet, more detailed statements concerning their true clinical value cannot be made for the time being as direct comparative studies of different techniques and systems are missing.

References

1. Wagnieres GA, Star WM, Wilson BC. (1998) *In vivo* fluorescence spectroscopy and imaging for oncological applications. *Photochem Photobiol* **68**: 603–632.

2. Chang SK. (2004) *Understanding the Variations in Fluorescence Spectra of Gynecologic Tissue.* Dissertation at the University of Texas in Austin, USA.

3. Betz CS, Stepp H, Janda P, Arbogast S, Grevers G, Baumgartner R, Leunig A. (2002) A comparative study of normal inspection, autofluorescence and 5-ALA-induced PPIX fluorescence for oral cancer diagnosis. *Int J Cancer* **97**: 245–252.

4. Arens C, Dreyer T, Glanz H, Malzahn K. (2004) Indirect autofluorescence laryngoscopy in the diagnosis of laryngeal cancer and its precursor lesions. *Eur Arch Otorhinolaryngol* **261**: 71–76.

5. Sankaranarayanan R, Ramadas K, Thomas G, Muwonge R, Thara S, Mathew B, Rajan B. (2005) Effect of screening on oral cancer mortality in Kerala, India: A cluster-randomised controlled trial. *Lancet* **365**: 1927–1933.

6. Kujan O, Glenny AM, Oliver RJ, Thakker N, Sloan P. (2006) Screening programmes for the early detection and prevention of oral cancer. *Cochrane Database Syst Rev* **3**: CD004150.

7. Driemel O, Kunkel M, Hullmann M, von EF, Muller-Richter U, Kosmehl H, Reichert TE. (2007) Diagnosis of oral squamous cell carcinoma and its precursor lesions. *J Dtsch Dermatol Ges* **5**: 1095–1100.

8. Lingen MW, Kalmar JR, Karrison T, Speight PM. (2008) Critical evaluation of diagnostic aids for the detection of oral cancer. *Oral Oncol* **44**: 10–22.

9. Patton LL, Epstein JB, Kerr AR. (2008) Adjunctive techniques for oral cancer examination and lesion diagnosis: A systematic review of the literature. *J Am Dent Assoc* **139**: 896–905.

10. Gray M, Gold L, Burls A, Elley K. (2000) The clinical effectiveness of toluidine blue dye as an adjunct to oral cancer screening in general dental practice. Report by the Department of Public Health and Epidemiology, University of Birmingham, UK.

11. Arens C, Glanz H, Dreyer T, Malzahn K. (2003) Compact endoscopy of the larynx. *Ann Otol Rhinol Laryngol* **112**: 113–119.

12. Watanabe A, Tsujie H, Taniguchi M, Hosokawa M, Fujita M, Sasaki S. (2006) Laryngoscopic detection of pharyngeal carcinoma in situ with narrowband imaging. *Laryngoscope* **116**: 650–654.

13. Ugumori T, Muto M, Hayashi R, Hayashi T, Kishimoto S. (2009) Prospective study of early detection of pharyngeal superficial carcinoma with the narrowband imaging laryngoscope. *Head Neck* **31**: 189–194.

14. Mehlmann M, Betz CS, Stepp H, Arbogast S, Baumgartner R, Grevers G, Leunig A. (1999) Fluorescence staining of laryngeal neoplasms after topical application of 5-aminolevulinic acid: Preliminary results. *Lasers Surg Med* **25**: 414–420.

15. Leunig A, Betz CS, Mehlmann M, Stepp H, Arbogast S, Grevers G, Baumgartner R. (2000) Detection of squamous cell carcinoma of the oral cavity by imaging 5-aminolevulinic acid-induced protoporphyrin IX fluorescence. *Laryngoscope* **110**: 78–83.

16. Andersson-Engels S, Klinteberg C, Svanberg K, Svanberg S. (1997) *In vivo* fluorescence imaging for tissue diagnostics. *Phys Med Biol* **42**: 815–824.

17. Malzahn K, Dreyer T, Glanz H, Arens C. (2002) Autofluorescence endoscopy in the diagnosis of early laryngeal cancer and its precursor lesions. *Laryngoscope* **112**: 488–493.

18. Paczona R, Temam S, Janot F, Marandas P, Luboinski B. (2003) Autofluorescence videoendoscopy for photodiagnosis of head and neck squamous cell carcinoma. *Eur Arch Otorhinolaryngol* **260**: 544–548.

19. Sharwani A, Jerjes W, Salih V, MacRobert AJ, El Maaytah M, Khalil HS, Hopper C. (2006) Fluorescence spectroscopy combined with 5-aminolevulinic acid-induced protoporphyrin IX fluorescence in detecting oral premalignancy. *J Photochem Photobiol B* **83**: 27–33.

20. Fielding D, Agnew J, Wright D, Hodge R. (2006) DAFE autofluorescence assessment of oral cavity, larynx and bronchus in head and neck cancer patients. *Photodiagnosis and Photodynamic Therapy* **3**: 259–265.

21. Morawiec-Sztandera A, Zimmer-Nowicka J, Kaczmarczyk D, Niedzwiecka I. (2008) Use of tissue autofluorescence in the diagnostics and assessment of treatment efficacy of the head and neck cancer. *Otolaryngol Pol* **62**: 540–544.

22. Kulapaditharom B, Boonkitticharoen V. (1998) Laser-induced fluorescence imaging in localization of head and neck cancers. *Ann Otol Rhinol Laryngol* **107**: 241–246.

23. Lane PM, Gilhuly T, Whitehead P, Zeng H, Poh CF, Ng S, Williams PM, Zhang L, Rosin MP, Macaulay CE. (2006) Simple device for the direct visualization of oral-cavity tissue fluorescence. *J Biomed Opt* **11**: 024006.

24. Poh CF, Zhang L, Anderson DW, Durham JS, Williams PM, Priddy RW, Berean KW, Ng S, Tseng OL, MacAulay C, Rosin MP. (2006) Fluorescence visualization detection of field alterations in tumor margins of oral cancer patients. *Clin Cancer Res* **12**: 6716–6722.

25. Huff K, Stark PC, Solomon LW. (2009) Sensitivity of direct tissue fluorescence visualization in screening for oral premalignant lesions in general practice. *Gen Dent* **57**: 34–38.

Chapter 5

Photodynamic Diagnosis and Photodynamic Therapy Techniques

*Zheng Huang**

Historical Background

Natural light has been used in the detection and treatment of disease for a very long time. However, the ability of visible light to destroy living organisms in the presence of photosensitizer was observed just a little over 100 years ago. As a medical student Oscar Raab worked with Professor Hermann von Tappeiner on the effect of acridine orange on the behavior of paramecia in the Pharmacological Laboratory of Munich Dermatology Clinic (Germany) and found that the acridine dye only killed the paramecia in the presence of light.[1] von Tappeiner *et al.* later recognized that oxygen was essential for generating the photocytotoxicity and he introduced the term *"Photodynamische Wirkung"* to describe the photosensitization phenomenon.[2] Thus was coined the term *"Photodynamic Therapy"*.

Hematoporphyrin (an iron-free derivative of heme) was first produced in 1841 by Johann Scherer (Germany) during his investigations of human blood. It was not until 1867 that the fluorescent properties of hematoporphyrin were revealed with the name "Hp" being given to the compound in 1871. However, modern photodynamic diagnosis (PDD) and photodynamic therapy (PDT) only

* Department of Radiation Oncology, University of Colorado Denver, Aurora, Colorado 80045, USA. MOE Key Laboratory of OptoElectronic Science and Technology for Medicine, Fujian Normal University, Fuzhou, P. R. China.

began to form in the 1960s after Richard Lipson and Edward Baldes at the Mayo Clinic (USA) reported that neoplastic tissues containing Hp would fluoresce under light irradiation.[3] Although early studies demonstrated the potential of Hp as a fluorescent marker for localizing cancers, one limitation was a high dose required to achieve consistent photosensitizer uptake in cancers, which could cause undesired cutaneous photosensitivity. The porphyrin mixture, prepared by Samuel Schwartz (Mayo Clinic), and formally named as hematoporphyrin derivative (HpD), was found to have a better affinity for cancerous cells, more intense fluorescence and stronger photocytotoxicity than crude Hp.[4]

Early studies in the 1970s were quickly expanded to investigate the optical properties and photodynamic effects of various HpD. Ivan Diamond *et al.* (USA) demonstrated that the combination of HpD and white light could result in regression of glioma in an animal model. An early clinical trial at Roswell Park Cancer Institute (USA) suggested that HpD-mediated PDT could result in the complete remission of various malignant tumors.[5] An effort led by Thomas Dougherty (Roswell Park Cancer Institute) to prepare a drug grade HpD produced the first approved photosensitizer agent, Photofrin®, for tumor ablation in the early 1990s in Canada. Subsequently, several light applicators were also developed to facilitate various PDT and PDD protocols. Regulatory approvals for the clinical use of several photosensitizers and light applicators for the treatment of non-malignant and malignant diseases now exist in many countries.[6–8]

Photosensitizers

The photosensitizer is a critical element in PDD and PDT. In general, the photosensitizer can be divided into three families based on its origin and structure:

  (i) *porphyrin-based photosensitizers* (e.g. Photofrin, ALA/PpIX, BPD-MA),
 (ii) *chlorophyll-based photosensitizers* (e.g. bacteriopherophorbides, plant chlorophyll A, chlorins, purpurins), and
(iii) *dyes* (e.g. phthalocyanine series, phenothiazinium series).

Since HpD, plant and bacterial chlorophylls, purpurins and phthalocyanine derivatives consist of a similar tetrapyrrole ring — a common structure of porphyrins, photosensitizers are also classified as porphyrins and non-porphyrins.[9]

Traditionally, those photosensitizers developed in the 1970s and early 1980s (e.g. Photofrin) are called first generation photosensitizers. Porphyrin derivatives, chemically pure agents or synthetics (e.g. mTHPC) made since the late 1980s are called second generation photosensitizers. Third generation photosensitizers generally refer to the modifications bound to carriers such as a biological conjugate (e.g. antibody conjugate) and built-in photon quenching or bleaching capability (e.g. PDT molecular beacon).[10] These terms are still being used, although not accepted unanimously since dividing photosensitizers into such generations may be confusing. In many cases, the claim that newer generation drugs are better than older ones is unjustified.[11] The premature conclusions on some investigational photosensitizers may send a misleading message by suggesting that the older drugs should be replaced by the newer ones or wrongly imply that newer photosensitizing drugs are superior to older ones.

Nevertheless, an ideal photosensitizer should meet some of the following criteria that are clinically relevant:

- pure chemical or precursor molecule (prodrug),
- low dark toxicity but strong photocytotoxicity,
- good selectivity towards targeted cells,
- high one- or two- photon cross section at desired wavelength(s),
- strong and distinct fluorescence emission (for PDD purposes), rapid removal from the body, and
- ease of delivery through various administration routes in a monomeric form.

Although some photosensitizers satisfy some or all of these criteria, there are currently only a few PDD and PDT photosensitizers that have received national or regional regulatory approval or are under clinical trials (see Table 1).

Table 1. Regulatory status of some common photosensitizers.

Photosensitizer	Abbreviation	Generic name	Manufacturer
Approved PDD photosensitizer			
Hexyl aminolevulinate	HAL	Hexvix	PhotoCure ASA
Approved PDT photosensitizers			
Polyhematoporphyrin ether/ester	Porfimer sodium	Photofrin	Axcan Pharma, Inc.
Hematoporphyrin derivative	HpD	Photogem	Moscow Institute of High Chemical Technologies
Benzoporphyrin derivative monoacid ring A	BPD-MA, verteporfin	Visudyne	Novartis Pharmaceuticals
5-aminolevulinic acid	ALA	Levulan	DUSA Pharmaceuticals, Inc.
Methyl aminolevulinate	MLA	Metvix	PhotoCure ASA
Meta-tetrahydroxyphenylchlorin in liposomes	mTHPC, temoporfin	Foscan	Biolitec AG
Mono-L-aspartyl chlorin e6 or talaporfin sodium[a]	NPe6, ME2906	Laserphyrin	Meiji Seika Kaisha, Ltd.
Mixture of chlorin e6, chlorin p6, purpurins 7 and 18		Photodithazine	VETA-GRAND Co.
Chlorin e6 in low-molecular polyvinylpyrrolidone	Fotolon	Photolon	RUE Belmedpreparaty
Sulfonated aluminum phthalocyanine	$AlPcS_{2-4}$	Photosens	General Physics Institute
Tolonium chloride or Toluidine Blue O	TBO	SaveDent PAD System	Denfotex Ltd.

(Continued)

Table 1. (*Continued*)

Photosensitizer	Abbreviation	Generic name	Manufacturer
Under clinical trials			
Lutetium(III) texaphyrin or motexafin lutetium	Lutex	Antrin	Pharmacyclics Inc.
Tin ethyl etiopurpurin	SnET2, purlytin	Photrex	Miravant Medical Technologies
Hematoporphyrin monomethyl ether	HMME	Hemoporfin	Fudanzhangjiang BioPharmaceutical Co., Ltd.
2-[1-Hexyloxyethyl]-2-devinyl pyropheophorbide-a	HPPH	Photochlor	Roswell Park Cancer Institute
Sodium salt of chlorin e6	Chlorin e6	Radachlorin	RADA-PHARMA
N-dimethylglucamine salt of chlorin e6		Fotoditazin	VETA-GRAND Co.
Pd-bacteriopheophorbide	WST09	Tookad	Negma-Lerads and Steba Laboratories Ltd.
Pd-bacteriopheophorbide monolysine taurine	WST11	Stakel	Negma-Lerads and Steba Laboratories Ltd.

[a] Also under clinical trials under different names: LS11 or Litx (Light Science Co.).

Note: The chemical structures of these photosensitizers can be found in Refs. 7, 9 and 12.

In general, the delivery process in PDD and PDT can be divided into "passive" and "active" based on whether a targeting component is presence or absence on the photosensitizing molecules.[12,13] Either way, the ideal delivery system should enable the selective accumulation of the photosensitizer within the target tissue and with little by non-target tissue. The drug carrier must be able to incorporate the photosensitizer without loss photosensitizer's activity. Liposomes, oil dispersions or micelle systems, biodegradable polymeric particles and hydrophilic polymer have often been used to enhance passive diffusion and phagocytosis processes. To provide selective and specific delivery of photosensitizer to target cells, a receptor-targeting moiety approach can be used to make an active targeted photosensitizer. Lipoprotein-mediated delivery and monoclonal antibodies have been used to reach these goals. As an emerging paradigm, nanoparticles have also been used as passive carriers or active components in the fluorescence and photodynamic processes.[14]

Photodynamic Diagnosis Techniques

PDD mechanisms

The interactions between photon and biological tissue and their detection are of great interest for many medical applications. Such interactions are dictated by a number of factors,[15] specifically:

- The nature of the incident light, which include the spectral parameters (wavelength, single or broad bandwidth), the power incident and density upon the target tissue.
- The optical properties of the target tissue, which are wavelength dependent and defined by the tissue scattering and absorption which are governed by the components and structures of target tissue.

In the ultraviolet and visible spectra, the tissue optical properties are dominated by intrinsic light absorbing molecules (i.e. chromophores). During the interaction between light and tissue, light absorption and energy transfer can initiate the process of re-emission of fluorescence light. In this chapter, "fluorescence" refers to the emission of visible light by a fluorescent substance upon being exposed to light irradiation, usually UV/visible wavelength. As long as the incident light continues to bombard the fluorescent substance, electrons in the substance become excited but return very quickly to lower energy by emitting light of longer wavelength which ceases as soon as the bombarding irradiation is discontinued. This phenomenon of fluorescence emission displays a number of general characteristics, such as excitation and emission spectra, the Stokes shift, quantum yields and lifetimes.[16]

The application of fluorescence examination (imaging and spectroscopy) for the characterization of biological materials and processes has been well established for several decades, based on the specific localization of fluorescent molecules (fluorophores) in cells and tissues. The *ex vivo* and *in vitro* application of these techniques is routinely performed in, for example, fluorescence microscopy and flow cytometry. Techniques frequently used in clinic *in vivo* include fluorescein angiography, fluorescence-assisted diagnosis, and fluorescence-guided resection. The fluorophores used in these imaging techniques can be considered in three main categories:

(i) endogenous (native) fluorophores (e.g. NADH, tryptophan) that are responsible for intrinsic tissue fluorescence (autofluorescence),

(ii) fluorophores synthesized by target cells after administration of a precursor molecule, specifically protoporphyrin IX (PpIX) induced by prodrug ALA or its esters, and

(iii) exogenous fluorophores administered as a contrast agent, including indocyanin green (ICG) and fluorescent photosensitizers used in PDT.

Fluorescence-assisted diagnosis is of increasing interest for diagnosis in dermatology and oncology. It is based on a more intense incorporation of a fluorescent dye in target lesions compared to normal tissue. Traditionally, the process of fluorescence-assisted diagnosis with the assistance of administrating a fluorescent PDT photosensitizer or prodrug is often called "photodynamic diagnosis (PDD)" although the process is a genetically related partner of PDT but does not actually involve the photodynamic reaction.

PDD application

As one unique form of photodiagnosis (PD), PDD may be used:

  (i) to differentiate benign versus malignant lesion,
 (ii) to assist in demarcating normal from diseased tissue, and
(iii) to assess biological responses to a treatment.

Different contrast agents and optical tools are required to complete these processes although some of which may not be readily available. Ideally, the optimal tool for PDD applications should be reliably and quickly undertaken. This tool also should be highly mobile, light-weight, and durable to allow for use both in the hospital setting and in the field as a screening tool. It should be emphasized that the successful combination of agent and device should offer high sensitivity and specificity.[17]

Photodynamic Therapy Techniques

PDT mechanisms

In addition to the re-emission of light as described in previous section, upon capturing light energy the excited photosensitizer can also undergo Type I and Type II photochemical reactions via the intersystem crossing process.[18] In a Type I reaction, the excited photosensitizer can transfer a proton or an electron to certain substrates to form a radical anion or radical cation. These radicals may further interact with oxygen to produce oxygenated products. In a Type II reaction, the excited photosensitizer can transfer its energy directly to molecular oxygen to generate singlet oxygen ($^1O_2$). Singlet oxygen is highly destructive because it reacts with most organic molecules at almost diffusion controlled rates. It is involved in addition reactions to enes and dienes, forming hydroperoxides and endoperoxides which can further propagate free radical chain reactions.[19] The generation of cytotoxic reactive oxygen species (ROS) can subsequently induce various biological and therapeutic effects.

Biological effect of PDT

Direct cytotoxicity

Since cytotoxic species have a short lifetime and only act within a limited distance, the uptake and subcellular localization of photosensitizer by tumor or other cells is critical in PDT-mediated direct cytotoxicities. The subcellular localization is determined by the chemical properties, formulation, concentration and delivery route of the photosensitizing drug, the microenvironment of the lesion, and to a certain degree the phenotype of the target cell. The plasma membrane, intracellular membranes and organelles such as the endoplasmic reticulum, Golgi apparatus, lysosomes, mitochondria and nucleus, have been identified as subcellular targets of photosensitizers.[12] Some photosensitizers distribute very broadly in these membranes and organelles. Some distribute specifically in the lysosomes or mitochondria.

The cell genotype, photosensitizer subcellular location and the drug/light dose might determine whether cell death occurs via apoptosis, autophagy or necrosis. In general, the mode of cell death switches from apoptosis to necrosis with the increase of the intensity of the insult. Interestingly, subthreshold dose can induce selective apoptosis of tumor cells, without detectable damage to normal cells. This might offer the possibility of treating patients with low-dose metronomic PDT to minimize damages to the normal tissue.[20]

Understanding PDT-induced unique changes and molecular events in signal transduction pathways, transcription factors and regulating mechanisms may provide a means to modulate and enhance PDT effects at the molecular level. In recent years there has been a steady increase in knowledge about the roles of the intracellular signaling machinery in cell response to PDT and, particularly, apoptotic pathways in PDT.[21–23] Two major apoptotic pathways have been

characterized: the death receptor-mediated (extrinsic pathway) and
the mitochondria-mediated apoptosis (intrinsic pathway). In general, experimental evidence shows a large heterogeneity in the
mechanisms leading to cell death in cellular-targeted PDT. However,
in addition to the activation of the molecular machinery leading to
cell death, PDT may also initiate metabolic reactions that protect
cells from oxidative stress. Therefore, the control of these protective
mechanisms is likely to enhance cytotoxicity on target cells.[24]

Advances in molecular biology allow a better understanding of
subcellular pathways and consequences of PDT cytotoxicity.
Although such advances help the rational design and choice of photosensitizers, the gold standard of evaluation of the efficacy of clinical
PDT still relies on the gross response of the lesion to PDT. For
instance, in tumor ablation, the clinically relevant response or biological endpoint would be acute tumor necrosis, the duration and
rate of this response, and the length of tumor-free period. These
might be affected not only by the direct cytotoxicity but also other
PDT effects as well as the tumor sensitivity to various PDT effects.

Vascular effect

It has been recognized since the early 1980s that an additional indirect mechanism, coexisting with the primary direct cytotoxicity of
antitumor PDT, is the vascular effect in which vascular damage
causes ischemic death and therefore provides another strategy to
treat solid tumors.[25] The mechanisms underlying vascular effects differ greatly with different photosensitizers, PDT protocols, and target
tissues.

Irradiation of photosensitizers, either confined in the blood circulation or accumulated in endothelial cells or bound to the vessel
walls, results in collateral damage to endothelial cells and leads to
the loss of tight junctions between cells and exposure of vascular
basement membranes. This primary damage within the vessel
lumen leads to the formation of thrombogenic sites and initiation of
cascade reactions such as platelet aggregation, the release of vasoac-
tive molecules, leukocyte adhesion, increase in vascular permeability

and vessel constriction. Microvascular collapse, blood flow stasis and tissue hemorrhages can lead to persistent post-PDT tumor hypoxia and long-term tumor control.[26–28]

Photosensitizers bound to carrier molecules, such as albumin, HDL or LDL, appear to have an active affinity to endothelial cells and tumor microvascular endothelium because of the existence of specific receptors in high numbers in these structures.[29] This might be an additive factor in photosensitizer uptake and retention in tumor tissue. Nevertheless, vascular-targeted PDT has certainly extended the application of PDT from ablating tumors to treating vascular lesions.[30,31]

Although one can expect that the localization of photosensitizer in both vascular and target cell compartments might produce a stronger combination effect, tumor response to vascular-targeted PDT does not correlate with the cellular concentration of photosensitizer in the treated tumor.[32] In vascular-targeted PDT mode therefore, the light irradiation should be applied during photosensitizer infusion and continued after the completion of infusion so as to center the light delivery period around the peak of the plasma concentration of photosensitizer and hence, putatively, should be maximizing the effect.[33]

Immune responses

Photodynamic therapy induced immune responses and particularly antitumor-specific immunity have been studied in various animal models.[7,28] Substantial evidence has demonstrated that antitumor-specific immunity and enhancement of host immune system might play important roles in secondary cytotoxicity and long-term tumor control to PDT, although these effects are not necessarily lethal to all tumor cells or relevant to the initial tumor ablation.[34–36]

Animal studies show that pro-inflammatory damages formed in cellular membranes and the blood vessel walls of treated sites start to recruit neutrophils, mast cells and monocytes/macrophages after PDT. These cells can also release more inflammatory mediators to enable massive recruitment of immune cells to tumor site.[37] These immune cells and nonspecific immune effector cells have a profound impact on PDT-mediated destruction of tumors. PDT can also activate the expression and production of several cytokines that regulate host immune response involving both lymphoid and non-lymphoid cells. PDT-induced uptake and presentation of tumor antigens by antigen presenting cells (APC) could initiate lymphocyte involvement.[38]

Complement-activating agents may further enhance the antitumor effect of PDT.[39] Recent studies also demonstrate that PDT-treated tumor cell lysates can be effective tumor vaccines, although the mechanism for enhancement of host anti-tumor immune responses to PDT vaccines is still unclear.[40] These advances in understanding PDT-induced immune responses might lead to an attempt to optimize PDT-mediated antitumor modalities through the modulation of important inflammatory/immune mediators.[41] Furthermore, combination strategies designed to activate host immunological stimulation (e.g. using immunoadjuvant) might further enhance the host immune responses and improve long-term tumor control.[42,43]

PDT light applicator and delivery

The first light sources used in PDT were non-coherent light sources (e.g. arc lamps). They can produce a wide spectrum of wavelengths to accommodate different photosensitizers. Although they are safer, easy to use, and less expensive, their disadvantages include significant thermal effect, low light intensity and difficulty in controlling and calculating the light dose. Light emitting diodes (LED) constitute another emerging PDT light applicator. LED can generate high energy light at desired wavelengths and can be assembled in different geometries and sizes. Needle-type probes can be implanted into solid tumors.[44,45] Large LED arrays, flexible organic LED film and textile-based light diffusers may be more suitable for the surface illumination of a wide-area superficial lesion.

Lasers are the most commonly used light source.[46] They produce high energy monochromatic light with a narrow bandwidth. Argon dye, potassium-titanyl-phosphate (KTP) dye, metal vapor lasers and, most recently, diode lasers, have been used in clinical PDT. The laser light can be focused, passed down an optical fiber and directly delivered to the target site through a specially designed illuminator tip. Optical fibers are compatible with self-expandable stents that are often used for luminal organs.[47] Four light delivery modes and their combination are commonly used in clinical settings:

(i) Front superficial irradiation — a uniform irradiance incident beam delivered to a surface by a microlens fiber externally,

(ii) Cavity superficial irradiation — an isotropic source centered in an essentially spherical cavity for delivering light to the cavity surface,

(iii) Cylindrical superficial irradiation — a cylindrical diffuser source centered in an essentially cylindrical lumen, and

(iv) Cylindrical interstitial irradiation — a cylindrical diffuser source embedded in the target tissue.

In cavitary and intraluminal PDT, transparent balloon catheters and light diffusing media (e.g. intralipid) can be used to provide better fiber positioning and dosimetry.

PDT dosimetry

Analytical optical modeling is a commonly used optical dosimetry approach based on the assumption that PDT is a threshold process, in which a minimum light energy density must be absorbed by the localized photosensitizer in order to initiate desired biological effects. The tissue necrosis depth or volume can be predicted based on optical parameters such as energy fluence and light penetration depth. These key parameters can be measured directly or calculated from the tissue optical constants for different light delivery modes.[48] However, successful PDT requires a sufficient light dose, photosensitizer dose and oxygen to generate sufficient cytotoxic species throughout the target tissue. Therefore, the concept of PDT dosimetry has been extended from optical dosimetry to quantify the distribution of the light fluence rate and the tissue optical properties as well as the photosensitizer concentration and the tissue oxygenation for optimizing or maximizing photodynamic effect while minimizing the dose to proximal normal structures.

Although new development in instrumentation allows light fluence rate, photosensitizer fluorescence intensity, and changes in local oxygen profile to be measured *in situ* in real-time, the protocols employed for clinical trials do not always consider important factors relevant to the PDT response, especially the dimensions and optical properties of the target tissue. It is not surprising that clinical PDT largely based upon empirical dose escalation trials without much consideration of the individual variations amongst patients has drawn criticism. The implementation of real-time dosimetry and individualized treatment plan are still a technical challenge.[49] PDT dosimetry presently under active research can generally be categorized as:

(i) Direct dosimetry — uses $^1O_2$ as PDT dose metric. However, direct $^1O_2$ monitoring (e.g. 1270 nm luminescence) is unlikely to become a routine tool in clinical or even preclinical PDT unless there is a substantial reduction in cost and complexity of instrumentation.[50]

(ii) Biological dosimetry — uses measurable change(s) in tissue that is correlated to the direct result of PDT. Methods such as CT, MRI and bioluminescence imaging have been used to assess PDT-induced biological effects. But it is still unclear whether these imaging techniques could be used to predict tissue responses and their outcome.

(iii) Implicit dosimetry — uses an implicit surrogate which is indicative of response to PDT. This method is appealing since only one quantity needs to be measured, for example, fluorescence photobleaching (utilizing the principles of PDD).

(iv) Explicit dosimetry — measures light fluence, drug concentration and tissue oxygenation. It is proposed that the energy absorbed by the drug per unit tissue volume would be a predictor of biological response. Cumulative PDT dose, activated at a particular wavelength for the drug, can be expressed as the time integral of light fluence rate $\varphi(q,t)$, and absorption coefficient $\mu_{ap}(q,t)$ of the photosensitizer:

$$Dose_{PDT}(q,T) = \int_0^T \varphi(q,t)\mu_{ap}(q,t)dt$$

where T is the total light irradiation time, q is a generalized spatial coordinate.

Each class has its own intrinsic merits and limitations. At present, there is no single dose metric that can address all dosimetry problems. Choice of dosimetry depends on the type and location of tumor, type and characteristic of drug and light, desired accuracy and instrumentation complexity.[49]

Modes of PDT application

Cellular-targeted PDT

Conventional cellular-targeted PDT is characterized by a long drug-to-light interval (DLI), for example, 24–48 h, to allow target cells to uptake adequate photosensitizer after the intravenous (i.v.) or oral administration of photosensitizer (e.g. Photofrin®, Foscan®, ALA). It has been used primarily for localized superficial or endoluminal malignant and pre-malignant conditions due to their accessibility to the light that can be easily delivered topically or endoscopically. Some photosensitizers and prodrugs (e.g. ALA and its esters) can also be applied topically to skin, mucosa and infectious lesions or intravesically to the bladder. In which cases, the DLI can be shortened to a few hours. Recently, the progress in light applicator and real-time online monitoring allows PDT application to expand to interstitial treatment of solid tumors.[49]

Vascular-targeted PDT

Vascular-targeted PDT is of growing interest and probably by far the most successful PDT application. It is characterized by a very short DLI, typically 0–30 min after the completion of i.v. injection of photosensitizer (e.g. Visudyn®, Tookad®).[33] Under this unique approach, light irradiation takes place while the photosensitizers are still circulating in the vascular compartment and therefore cause direct vascular damages through the low-density lipoprotein receptor-mediated endocytosis pathways and lead to thrombosis and microvessel occlusion. Vascular-targeted PDT has been used primarily for the management of the neovascularization lesion (e.g. age-related macular degeneration, AMD)[31] and capillary malformations (e.g. port-wine stain birthmarks).[30] Recently, vascular-targeted PDT has been investigated for curative or palliative treatment of solid tumors (e.g. prostate cancer) by targeting the tumor vasculature.[33] The massive shutdown of pathological and normal vessels can deprive the supply of oxygen and nutrients and subsequently achieve tumor ablation.[29]

Extracorporeal PDT

Extracorporeal PDT, also known as extracorporeal photophoresis, is an *ex vivo* approach which involves a short incubation of the whole blood or blood products with a photosensitizer (e.g. Riboflavin®, TH9402) and *ex vivo* light irradiation at a shorter wavelength (e.g. 285–514 nm). This process may or may not require a photosensitizer extrusion step before and after light irradiation. Extracorporeal PDT has been used for the pathogen inactivation in blood transfusion and selective cell purging in graft-versus-host disease (GvHD) prevention.[8,42] Noticeably, cross-linking anti-Fas antibody combined with PDT could have an additive impact against the survival of $CD41^+CD81^+$ thymocytes through proapoptotic pathways.

Photochemical internalization

The photochemical internalization (PCI) (light-directed delivery or photodynamic delivery) is a PDT-based technology for the delivery of macromolecules into the cytosol and it can overcome some of the challenges recognized in drug delivery and cancer treatment. Unlike PDT, the photosensitizer used in PCI process has to be located in the endocytic vesicles of the targeted cells and will release endocytosed therapeutic macromolecules after a photochemically induced rupture of the endocytic vesicles, therefore bypass the degradation in lysosomes. This light-directed site-specific drug delivery induced by photodynamic reactions will take place in addition to the well described PDT effects. Preliminary studies show that PCI can facilitate intracellular delivery of a large variety of macromolecules that do not otherwise readily penetrate the cell, including Type I ribosome-inactivating proteins (RIPs), RIP-based immunotoxins, genes and some chemotherapeutic agents.[51]

Future Prospects

There is still a strong and increasing interest and research effort internationally focused on developing new PDD and PDT photosensitizers, exploring PDT mechanisms at molecular level, enhancing PDT efficacy with combined modality and evaluating potential clinical indications of PDD and PDT. Some new strategies currently under development might break in some fundamental way from conventional strategies.[52–55] For example, newly approved HAL blue-light cystoscopy has revolutionized bladder cancer detection and care.[56] The utility of the target-specific photosensitizers in developing multimodality agents, such as tumor-imaging and therapy, represents another new strategy.[57] Two-photon excitation (TPE) and the development of photosensitizers with a large two-photon cross section have also received increasing attention due to its potential application in two-photon PDT and in high resolution two-photon fluorescent microscopy and imaging.[52,16] TPE delivers light energy at a high peak power with a comparatively low average power more precisely to the target tissue or cell with a high degree of spatial specificity.[58]

The birth of a new journal titled *Photodiagnosis and Photodynamic Therapy*, now entering its 6th year, also highlights such interest in PDD and PDT. The number of scientific articles on PDD and PDT clinical applications as well as basic science steadily increases in both English and non-English literatures. Review articles on past work, new aspects and future applications have been published on a regular basis while new technology and promising applications continue to be discovered.

Each year several international conferences are held regularly which bring together these interests and research. Examples include the World Congress of International Photodynamic Association (IPA), annual symposiums at Photonics West organized by the International Society for Optical Engineering (SPIE), annual meeting of European Platform for Photodynamic Medicine (EPPM), and International Symposium on Photodynamic Diagnosis and Therapy in Clinical Practice (Brixen, Italy). PDD and PDT sessions could also be found in medical laser, biophotonics and photobiology meetings.

Although regulatory approvals for the clinical use of photosensitizers and light applicators in PDD and PDT now exist in many countries around the world, the total number of approved clinical indications is still limited. There is an urgent need for involvement from industries and research institutes to continue to launch clinical trials to evaluate applications of new PDD and PDT techniques in conjunction with, or as a replacement for, conventional approaches.

Acknowledgments

The author wishes to thank the US National Institutes of Health (Grant CA-43892) for support of this work.

References

1. Raab O. (1900) *Z Biol* **524**. (in Germany)
2. von Tappeiner H, Jodlbauer AU. (1904) *Deutsches Arch Klin Med* **427** (in Germany)
3. Lipson RL, Baldes EJ. (1960) *Arch Dermatol* **508**.
4. Dougherty TJ, Henderson BW, Schwartz S, Winkelman JW, Lipson RL. (1992) In *Photodynamic Therapy*, Henderson BW, Dougherty TJ. (eds.), Maurice Dekker, New York.
5. Dougherty TJ, Kaufman JE, Goldfarb A, Weishaupt KP, Boyle D, Mittleman A. (1978) *Cancer Res* **2628**.
6. Huang Z. (2005) *Technol Cancer Res Treat* **283**.

7. Huang Z, Liu H, Chen WR. (2006) In *Series on Biomaterials and Bioengineering in Fundamentals & Applications of Biophotonics in Dentistry*, Kishen A, Asundi A. (eds.), World Scientific Publishing Co. Singapore.

8. Huang Z, Li L, Wang H *et al.* (2009) *J Innov Opt Health Sci* **73**.

9. O'Connor AE, Gallagher WM, Byrne AT. (2009) *Photochem Photobiol* **1053**.

10. Moser JG. (1997) In *Photodynamic Tumor Therapy — 2nd & 3rd Generation Photosensitizers*, Moser JG. (ed.), Harwood Academic Publishers, London.

11. Allison RR, Downie GH, Cuenca R *et al.* (2004) *Photodiag Photodyn Therapy* **27**.

12. Castano AP, Nemidova TN, Hamblin MR. (2004) *Photodiag Photodyn Therapy* **279**.

13. Konan YN, Gurny R, Allémann E. (2002) *J Photochem Photobiol B* **89**.

14. Chatterjee DK, Fong LS, Zhang Y. (2008) *Adv Drug Deliv Rev* **1627**.

15. Moghissi K, Stringer MR, Dixon K. (2008) *Photodiag Photodyn Ther* **235**.

16. Lakowicz JR. (2006) In *Principles of Fluorescence Spectroscopy*, 6th Edition, Springer Science & Business Media, LLC, New York.

17. Allison RR, Sibata CH. (2008) *Photodiagn Photodyn Ther* **247**.

18. Phillips D. (1994) *Sci Progress* **295**.

19. Foote CS, Shook FC, Abakerli RB. (1984) *Methods Enzymol* **36**.

20. Bisland SK, Lilge L, Lin A, Rusnov R, Wilson BC. (2004) *Photochem Photobiol* **22**.

21. Castano AP, Nemidova TN, Hamblin MR. (2005) *Photodiag Photodyn Ther* **1**.

22. Nowis D, Makowski M, Stoklosa T *et al.* (2005) *Acta Biochimica Polonica* **339**.

23. Almeida RD, Manadas BJ, Carvalho AP, Duarte CB. (2004) *Biochim Biophys Acta* **59**.

24. Gomer CJ, Luna M, Ferrario A *et al.* (1996) *J Clin Laser Med Surg* **315**.

25. Star WM, Marijnissen JP, van den Berg-Blok AE, Reinhold HS. (1984) *Prog Clin Biol Res* **637**.

26. Krammer B. (2001) *Anticancer Res* **4271**.

27. Abels C. (2004) *Photochem Photobiol Sci* **765**.

28. Castano AP, Nemidova TN, Hamblin MR. (2005) *Photodiagn Photodyn Therapy* **91**.

29. Chen B, Pogue BW, Hoopes PJ, Hasan T. (2006) *Crit Rev Eukaryot Gene Expr* **279**.

30. Yuan KH, Li Q, Yu WL *et al.* (2008) *Photodiag Photodyn Ther* **50**.

31. Kaiser PK, Visudyne. In Occult CNV (VIO) Study Group. (2009) *Curr Med Res Opin* **1853**.
32. Dolmans DE, Kadambi A, Hill JS *et al.* (2002) *Cancer Res* **4289**.
33. Huang Z, Chen Q, Luck D *et al.* (2005) *Lasers Surg Med* **390**.
34. Canti G, De Simone A, Korbelik M. (2002) *Photochem Photobiol Sci* **79**.
35. van Duijnhoven FH, Aalbers RIJM, Rovers JP, Terpstra OT, Kuppen PJK. (2003) *Immunobiol* **105**.
36. Nowis D, Stoklosa T, Legat M *et al.* (2005) *Photodiag Photodyn Ther* **283**.
37. Korbelik M, Krosl G, Krosl J, Dougherty GJ. (1996) *Cancer Res* **5647**.
38. Abdel-Hady ES, Martin-Hirsch P, Duggan-Keen M *et al.* (2001) *Cancer Res* **192**.
39. Korbelik M, Sun J, Cecic I, Serrano K. (2004) *Photochem Photobiol Sci* **812**.
40. Gollnick SO, Vaughan L, Henderson BW. (2002) *Cancer Res* **1604**.
41. Castano AP, Mroz P, Hamblin MR. (2006) *Nat Rev Cancer* **535**.
42. Qiang YG, Yow CM, Huang Z. (2008) *Med Res Rev* **632**.
43. Sur BW, Nguyen P, Sun CH, Tromberg BJ, Nelson EL. (2008) *Photochem Photobiol* **1257**.
44. Schmidt MH, Bajic DM, Reichert II KW *et al.* (1996) *Neurosurgery* **552**.
45. Lustig RA, Vogl TJ, Fromm D *et al.* (2003) *Cancer* **1767**.
46. Mang TS. (2004) *Photodiag Photodyn Therapy* **43**.
47. Wang LW, Li LB, Li ZS *et al.* (2008) *Lasers Surg Med* **651**.
48. Grossweiner LI. (1997) *J Photochem Photobiol B Biol* **258**.
49. Huang Z, Xu H, Meyers AD *et al.* (2008) *Technol Cancer Res Treat* **309**.
50. Niedre MJ, Yu CS, Patterson MS, Wilson BC. (2005) *Br J Cancer* **298**.
51. Norum OJ, Giercksky KE, Berg K. (2009) *Photochem Photobiol Sci* **758**.
52. Wilson BC. (2008) In *Advances in Photodynamic Therapy: Basic, Translational, and Clinical*, Hamblin MR, and Mroz P. (eds.), Artech House Inc., Book News, Inc., Portland.
53. Josefsen LB, Boyle RW. (2008) *Br J Pharmacol* **1**.
54. Lovell JF, Liu TW, Chen J, Zheng G. (2010) *Chem Rev* **2839**.
55. Verhille M, Couleaud P, Vanderesse R *et al.* (2010). *Curr Med Chem* **3925**.
56. Stenzl A, Burger M, Fradet Y *et al.* (2010) *J Urol* **1907**.
57. Celli JP, Spring BQ, Rizvi I *et al.* (2010) *Chem Rev* **2795**.
58. Collins HA, Khurana M, Moriyama EH *et al.* (2008) *Nat Photonics* **420**.

Chapter 6

OCT Detection of Lung Cancer

S. Murgu[*,‡] *and M. Brenner*[*,†,§]

Rationale for Early and Accurate Detection of Airway Cancer

Lung cancer is a leading cause of cancer-related death in the world and accounts for more deaths than breast, colon and prostate cancer combined in the United States.[1] Most patients present with advanced disease, and overall five year survival is dismal.[2] Early diagnosis appears to offer a better prognosis and allow less invasive treatment options, which is especially important in patients with significant comorbidities. For the vast majority of patients with limited stage non-small-cell bronchogenic carcinoma, accurate and complete surgical resection is the only potentially curative approach. Furthermore, accurate detection, staging and assessment of operative margins is important both for cure, and to avoid either excessive or inadequate resection surgery. Local recurrence, in fact, is not uncommon and occurs in 5 to 20% of patients, depending on the pathologic staging of the tumor at the time of diagnosis. Fifty percent of all recurrences are diagnosed in the first 24 months after presumed curative treatment, demonstrating shortfalls of current diagnostic capabilities.

Lung cancers arise in the periphery of the lung or in the airways. While peripheral lung cancers comprise the majority of lung cancer

* University of California, Irvine Medical Center, Orange, CA 92868–3217, USA
† Beckman Laser Institute, University of California, Irvine, USA
‡ smurgu@uci.edu
§ mbrenner@uci.edu

sites of origin, airway cancers still represent a sizable fraction of primary lung cancers. At the present time, the major near-term potential role for optical coherence tomography (OCT) in lung cancer diagnosis is in airway based origin lesions, due to accessibility issues. Therefore, this chapter focuses exclusively on airway lung cancer OCT detection. Further into the future, as more advanced needle probe-based OCT systems are developed, OCT may gain a role in peripheral parenchymal lung tumor diagnostics.

Current evidence suggests that OCT will be one facet in a broad armamentarium for diagnosis and assessment of airway cancers, functioning as part of a multimodality imaging array. OCT must therefore be understood in the context of other airway cancer imaging and treatment assessment approaches. In this chapter, we present OCT airway cancer imaging capabilities, limitations, current role and potential, and briefly describe some other emerging technologies that may be employed in a complementary manner with OCT in the emerging multimodality imaging approach development, and we speculate on how they may interrelate with OCT imaging.

The argument for early lung cancer detection strategy is that if treating lung cancer early improves outcome, then effective screening methods for lung cancer are justified.[3] Previous sputum cytology screening programs were based around the finding that squamous cell carcinomas represented 70% of all lung cancers. Squamous cell cancers, however, now represent only 25–30% of all lung cancers (although it still represents more total cancer related mortality, than breast and colon cancers combined). These cancers are thought to be preceded by a progression of cellular atypia, reversible pre-epithelial proliferation and carcinoma *in situ*.[4] Early stage squamous cell airway cancer, defined as ≤1 cm in diameter and 3 mm deep is considered N0 (TNM staging classification) cancer and can potentially be cured with bronchoscopic treatment as an alternative for surgical resection.[4] The genesis of squamous cell lung cancer seems to require increased genomic instability and cell proliferation early in the transformation process. Cellular transitions, DNA aneuploidy and p53 gene over-expression during the transition from bronchial squamous metaplasia to advanced carcinoma have been demonstrated. These cellular changes may lead to the eventual disruption of basement membrane and extension of intraepithelial carcinoma.[5] However, these early morphologic changes occur below the threshold detection limits of conventional white light bronchoscopy (WLB) which has only a 30% sensitivity to detect early stage cancer in the central airways.[6] A significant percentage of patients with moderate and severe dysplasia will develop invasive carcinoma during a multi-step carcinogenesis transformation that might take three to four years.[3] Approximately 11% of subjects with moderate dysplasia and 19 to 46% with severe dysplasia subsequently develop invasive cancer.[7,8] Furthermore, the cumulative risk of developing lung cancer in a patient with severe dysplasia or carcinoma *in situ* (CIS) was found to be more than 50 percent at two years.[9]

In addition to early detection, better identification of tumor margins and subsequent improved open surgical or bronchoscopic procedures with curative intent should potentially lead to reduction in recurrence rates, improved survival, and reduction in inadvertent non-curative surgical procedures. Therefore, it is important to develop improved methods for (1) early detection of airway lung cancer, (2) accurate diagnosis, and (3) determination of location, borders, extent, and stage of disease. For all these purposes, high resolution imaging modalities are needed (Fig. 1).

Non-Bronchoscopic Imaging Methods for Airway Cancer Detection

Standard non-invasive methods for lung and airway anatomic imaging include chest radiography, computed tomography (CT) scanning, magnetic resonance imaging (MRI), positron emission tomography (PET) scanning, and ultrasound. Chest radiographs are inexpensive and have excellent depth of penetration since the entire thorax can be imaged. Standard chest radiographs, however, are two dimensional composite shadows of tissue structures, with maximal resolution on the order of millimeters. Digital imaging and subtraction techniques

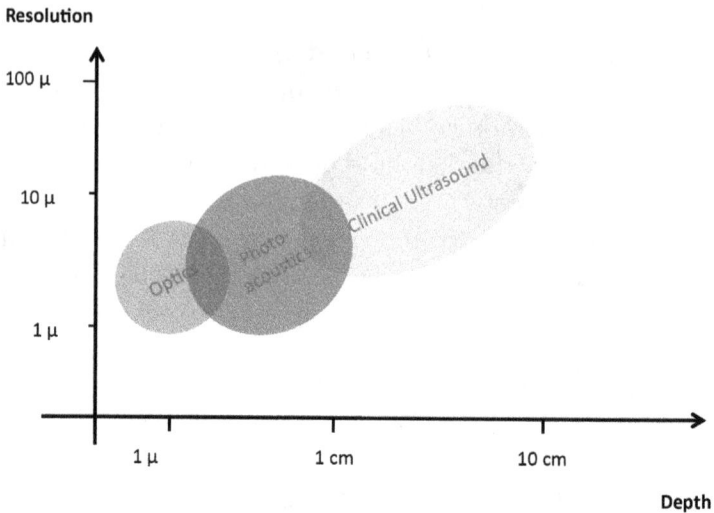

Fig. 1. Schematic representation of approximate resolution and depth of penetration of multimodality imaging technologies with potential for airway cancer diagnostics. As a general concept, systems applicable to airway cancer diagnostics with greater depth of penetration have lower resolution and cover larger regions, while systems with higher resolution have shallower depth of penetration and cover smaller regions. Optical technologies including confocal microscopy and optical coherence tomography offer micron level resolution with depth of penetration of <1–3 mm. Ultrasonography provides high penetration depths but only 50–200 μm resolution. Photoacoustics is not available for airway use at this time but covers the gap in penetration depth and resolution between optical and acoustic technologies and provides approximately 3–30 mm depth of penetration and a depth to resolution ratio of ~100.

may further improve diagnostic capabilities for chest radiographs and other imaging techniques.[10] High-resolution CT scanning is three-dimensional and also has excellent penetration with resolution on the order of hundreds of microns to millimeters.[11] Reconstruction algorithms can be applied and include airway surface renderings comprising "virtual" bronchoscopy".[12] However, virtual bronchoscopy is limited in ability to evaluate the mucosal surface of the respiratory tract reliably. Although the overall surface morphology can be detected, mucosal color, neovascularization, infiltration, or friability cannot be assessed.

MRI is used primarily for mediastinal and proximal vascular structure evaluation in the thorax at this time and occasionally in tracheobronchomalacia and for detecting involvement of the chest wall structures by tumor.[13,14] The depth of penetration is throughout the thorax and imaging resolution is on the order of millimeters, which is insufficient to resolve tissue microstructure.[11] Furthermore, high-speed image guidance which may be needed intra-operatively, for instance, is limited by size, cost, and complexity.[15]

PET scanning has relatively low resolution but provides functional information that may be useful in differentiating benign from malignant pathology and particularly in conjunction with CT may also be useful for detection of unsuspected metastatic malignant disease in lung cancer patients.[16] Transthoracic ultrasound is somewhat lower resolution in pulmonary applications that are designed to penetrate the chest wall. The most frequent applications have focused on assessment of pleural effusions, mediastinal and peribronchial adenopathy, or chest wall lesions but there is currently no defined role for airway pathology assessment.[17]

Conventional Bronchoscopic Imaging Methods for Airway Cancer Detection

Standard direct real time airway anatomic examination methods include flexible and rigid bronchoscopy. Flexible bronchoscopy is performed either trans-orally or trans-nasally using flexible endoscopes that are usually at least 100 cm in length, and have outer diameters generally ranging from 2.2 mm to 6.5 mm. The bronchoscopes most commonly have a light source, imaging modality (CCD chip or fiberbundle), and a working channel usually ranging from 1.2 to 2.8 mm. WLB can only image tissue surfaces. Specimens can be obtained via suctioning, saline lavage, brushing, needle and forceps biopsies (generally of maximal diameter on the order of 1–2.5 mm). Generally, 3–4 orders of bronchial branching can be visualized with standard flexible bronchoscopy. Ultrathin flexible bronchoscopy using flexible scopes of 2–3 mm and working channels of 1 mm allow access to more distal airways and pulmonary abnormalities peripherally situated within the lung parenchyma. A major advantage of flexible bronchoscopy is its versatility. Flexible bronchoscopy can be performed with or without moderate sedation, to relive patient's discomfort, pain and anxiety. Procedures can be performed using topical anesthesia, and bronchoscopy can be done at the patient's bedside or in an outpatient setting.[18]

Rigid bronchoscopy is usually performed in conjunction with interventional therapeutic procedures. These include laser ablation, tumor resection, airway stent insertion, or dilatation of airway strictures. Rigid bronchoscopes generally range in size from 5 to 12 mm diameter and 12 to 45 cm in length and have working channels as large as 7 mm in addition to the visualization optics. Rigid bronchoscopes can visualize up to 4 orders of bronchial branching and are primarily focused on proximal airways. Airway surface is visualized using straight lens telescopes with a digital or other high grade video camera easily attached to the telescope head.[18] The probes or catheters placed through the rigid tube need to be sufficiently firm so that they can be manipulated inside the open tube. This is the case, for example, for existing silicone suction catheters, quartz laser delivery fibers, high frequency endobronchial ultrasound or OCT probes (Fig. 2). These probes could be manipulated through the open tube system and do not necessarily need to be placed into a secured guidance system within the rigid bronchoscope itself.

Limitations of Conventional White Light Bronchoscopy for Airway Cancer Detection

Conventional WLB is not useful for detecting mucosal changes that might be just a few cells thick or below the tissue surface. Furthermore, as the number of branching airways increases, their individual luminal diameters decrease in size and increase exponentially in number.[19] Therefore, it is not possible to examine most of the conducting airways within a patient (in contrast for example, to the single lumen found in the gastrointestinal tract). Only a few generations (~5) of branching can be routinely examined by standard bronchoscopic diagnostic techniques. Despite these limitations, a significant amount of airway cancer

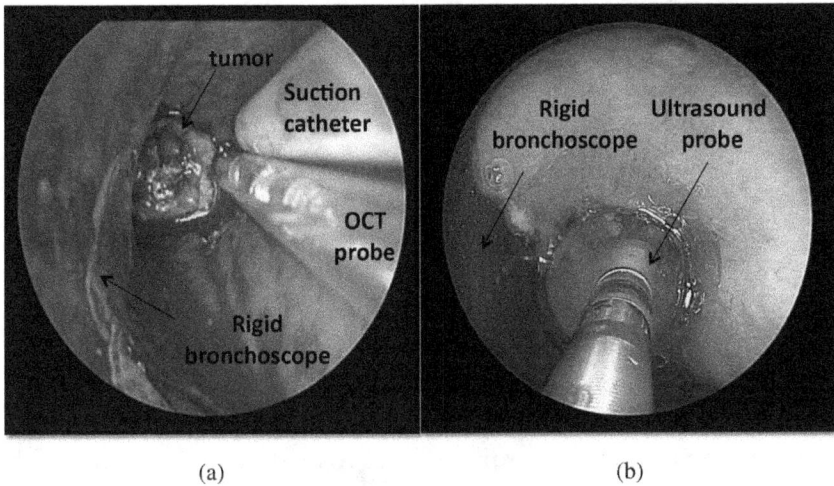

Fig. 2. (a) A transverse scanning OCT prototype probe and suction catheter are seen inside a 12 mm rigid bronchoscope overlying a left lower lobe malignant bronchial tumor. (b) A high frequency endobronchial ultrasound radial probe is seen overlying the rigid bronchoscope and the airway wall.

occurs in the proximal airways, at levels reachable by current bronchoscopic methods and within areas that could potentially be visualized by application of more advanced optical technologies.

In diagnosis and treatment of airway lesions, selection of biopsy sites is prone to sampling errors and biopsies may be associated with complications including bleeding and pneumothorax or persistent air-leak. This limits the number of biopsies and potential sampling sites, as well as diagnostic accuracy, while it increases risks and expense. Biopsy yield rates vary based on the underlying disease, the patient's ability to tolerate the procedure, and for malignant disease — on the location, depth, appearance, size, extent, cell type, degrees of vascularity and necrosis.[18]

Endobronchial tumors often have regions of fibrosis, neovascularity, necrosis, or areas where the tumor is located below the visible airway surface, leading to non-diagnostic endobronchial biopsies or misinterpretation of malignancy borders. Using current WLB, the bronchoscopic yield in terms of diagnostic sensitivity and specificity for endobronchial malignancy is variable. Rates for diagnosing visible lesions in patients with endobronchial tumors have been reported between 75–98%. Overall specificity has been reported at 64–96% and pathologic diagnostic error rates of 1.5–3%. For central tumors the maximal diagnostic yield was not achieved until the 4th forceps biopsy.[20] Improved guidance for airway biopsy would therefore increase yield and decrease risks associated with the large number of biopsies currently necessary for diagnosis. This is potentially offered by newly developed radiographic, acoustic and optical technologies.

Disadvantages of rigid bronchoscopic approaches include the need for general anesthesia, relatively limited clinical indications, high levels of operator skills needed, and ability to view only more proximal order airway branches compared to flexible bronchoscopy. Operator experience allows procedures to be done quickly and safely in both the inpatient and outpatient setting with minimal recovery times based primarily on the amount of sedation employed.[18]

Alternative and Complementary Bronchoscopic Imaging Modalities for Airway Cancer Detection

Autofluorescence bronchoscopy

Fluorescence bronchoscopy is based on the principle that differences in epithelial thickness, blood flow, and tissue fluorophore concentration cause abnormal tissues to have diminished red and substantially diminished green fluorescence as compared to normal tissues when exposed to 440–480 nm wavelength blue light excitation.[3] However, this autofluorescence (AF) cannot be seen during conventional WLB as the intensity of fluorescence is very low, and is overwhelmed by the reflected and back-scattered light. The major fluorescing chromophores in the airway mucosa are likely elastin, collagen, flavins, nicotinamide-adeninedinucleotide (NAD), NADH (hydrogen), and porphyrins. Accelerated intracellular metabolism in cancer cells decreases riboflavin and flavin coenzymes and NADH caused by overproduction of lactic acid through glycolysis, leading to the autofluorescence changes seen during the progression to malignant transformation.[21]

Because intraepithelial lesions are only a few cell layers thick, surface mucosa is typically normal appearing during conventional WLB and only a third of CIS are visible to the experienced bronchoscopist.[6] The white light bronchoscopy appearance of CIS can look similar to those produced from chronic bronchitis, which is often present in patients at risk for lung cancer. By the time the mucosa appears distinctly abnormal, the cancer is typically invasive. As mucosal and submucosal disease progresses from normal, to metaplasia, dysplasia and CIS, there is a progressive loss of the green AF, causing a red-brown appearance of the tissue. Areas of abnormal fluorescence are identified as brown regions on a normal green background. Successful localization of a dysplastic lesion is possible using AF while white light examination is normal. The intensity of green fluorescence is measured and an abnormally low "green" fluorescent area is identified as a cold spot on a normal green background (Fig. 3). Detection of abnormal tissues and discrimination from normal bronchial mucosa using this technology continues to improve. In a recent study, violet light excitation at 405 nm was delivered (rather than the 450 nm blue wavelength), and the contrast between pre-neoplastic and healthy tissue was quantified using off-line image analysis. There was almost three times higher contrast when backscattered red light was added to the violet excitation.[22] AF has been shown to be helpful in high risk groups such as in patients with a history of head and neck cancer, COPD, and smokers. Roentgenographically occult lung cancers are characterized by an incidence of synchronous lesions ranging from 0.7–14%, and that metachronous lesions might occur in as many as 5% per year.[23,24] AF bronchoscopy has therefore been advocated as a surveillance tool in these patients as well as in patients with newly diagnosed early cases before thoracotomy, and in those patients who have undergone curative surgery for non-small cell lung cancer.[25] An increased sensitivity to about 86% of WLB combined with AF has been demonstrated.[6] However, lack of specificity has limited the overall role of AF bronchoscopy in airway cancer detection. Additional concerns are that in light of the low specificity, more numerous biopsies are actually taken, reducing the overall yield per biopsy.[26]

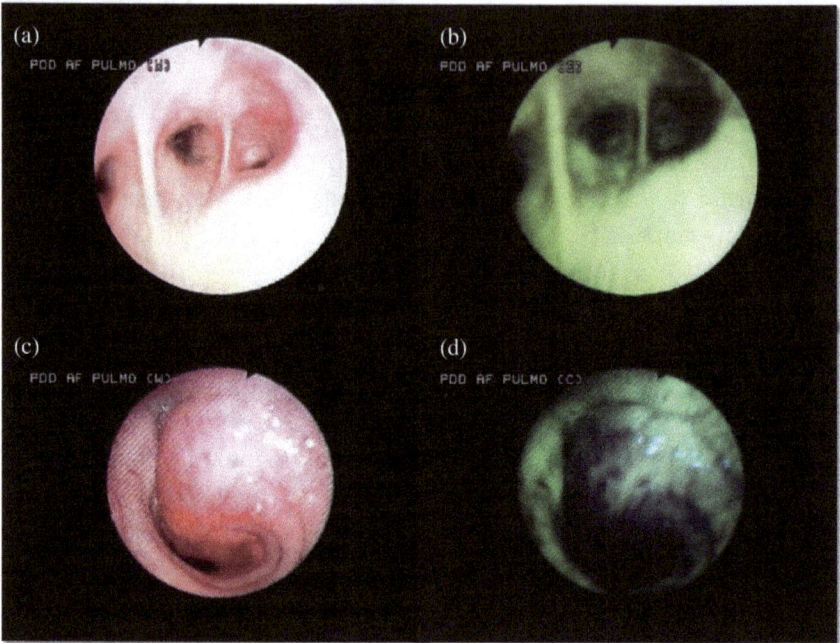

Fig. 3. (a) Normal left lower lobe airway mucosa seen during white light bronchoscopy. (b) The corresponding normal green fluorescence of the airway is seen during AF bronchoscopy. (c) Tracheal tumor with neovascularization and abnormal mucosa is visualized during white light bronchoscopy. (d) The corresponding abnormal fluorescence is visualized as brown-purple region on a background of normal green fluorescence during AF bronchoscopy.

High magnification bronchoscopy

The only abnormality usually detected on standard WLB in dysplasia is edema and erythema at bronchial bifurcations. However, histologically there is neovascularization or prominent vascularity. High magnification bronchoscopy (HMB) enables observation of vascular networks in order to identify potential areas of increased vascularity in the airway wall in patients with lung cancer.[27] This system can provide information on the airway mucosa with a maximum magnification of 110 times. A dense concentration of subepithelial microvessels is observed mainly in the inter cartilage portion, indicating an increase in submucosal circulation, making it a useful tool for observing and evaluating the subepithelial microvessels in large airways in various pathologies such as asthma or lung cancer.[3]

Narrow band imaging

Microvascular structures are further resolved if recently described narrow band filter techniques are used instead of the conventional RGB broadband filter during high magnification bronchoscopy. The narrow band imaging (NBI) technique uses a 415 nm blue light which is absorbed by hemoglobin contained in the capillary network on the mucosal surface and a 540 nm green light that is absorbed by blood vessels located a bit deeper below the capillary layer. Recently, in a study evaluating early lung cancer detection, the sensitivities of differentiating between CIS/severe dysplasia and mild or moderate dysplasia or metaplasia by AF were 83% and 52% compared to 100% and 90% by NBI, respectively. Thus, the specificity of NBI (90%) was significantly higher than that of AF (52%).[28] When compared with WLB alone, NBI increased the rate of detection of dysplasia or malignancy by 23%.[28,29]

Endobronchial ultrasound

Ultrasound and Doppler ultrasound image tissue structure and blood flow, but are limited in spatial resolution to approximately 50–200 μm due to their relatively long acoustic wavelengths.[30] However, high resolution endobronchial ultrasound (EBUS) with radial scanning 20 MHz transducers has been used to accurately measure depth of tumor invasion beyond the cartilaginous layer and to identify the structural layers of the airway wall that are important in defining and understanding various central airway disorders (Table 1).

For lung cancer, this technology provides important information since it is known that lymph node invasion in radio-occult cancers changes staging, treatment, and prognosis. Although radio-occult lymph node positivity is very rare in CIS (1%) (and is not seen in lesions < 3 mm thick, 10 mm in surface area, or with an invasion index < 20 mm in the large central bronchi), significant rates (10 to 30%)

Table 1. Reported high-resolution optical and acoustic technologies with potential for multimodality airway cancer imaging roles.

Technology	Spatial (axial) resolution	Scanning depth	Lateral (transverse) resolution	Image acquisition time/mode	Image orientation	*In vivo* pulmonary-airway applications to date
Optical coherence tomography-time domain	4–20 μm	2–3 mm	21–27 μm	~15 frames/s Non contact	Cross section	yes
Fourier Domain OCT-Optical frequency domain imaging	7 μm	1–2 mm	15 μm	10–100 frames/s Non contact	Cross section	no
Confocal microscopy	0.7–3.5 μm	0.5 mm	1–15 μm	12 frames/s contact	En face	yes
Photoacoustics	Depth/resolution ratio ~100	3–30 mm	n/a	25 seconds contact	Cross section	no
High-resolution endobronchial ultrasound	50–200 μm	~50 mm	n/a	30 seconds contact	Cross section	yes

of lymph node invasion have been reported in case of invasion of the cartilage.[3] Furthermore, 70% of patients presenting with radiographically occult lung cancer were revealed to have more advanced cancer by combination of AF bronchoscopy and high-resolution CT than by the initial evaluation.[31] Thus, accurate evaluation of peribronchial tumor invasion is important prior to endoscopic or surgical treatment. For this purpose, the high frequency radial endobronchial ultrasound probe has been proven to be useful.[32]

The standard frequency for radial probe EBUS is 20 MHz, which allows for a resolution of less than 1 mm, and a depth of penetration of approximately 4 to 5 cm, allowing visualization of mediastinal structures. This frequency allows for assessment of the layers of the airway wall and the peribronchial structures (Fig. 4).[33] EBUS clearly defines the bronchial wall layers and adjacent anatomic structures and has demonstrated effectiveness for distinguishing airway tumor invasion from external tumor induced airway compression, as well

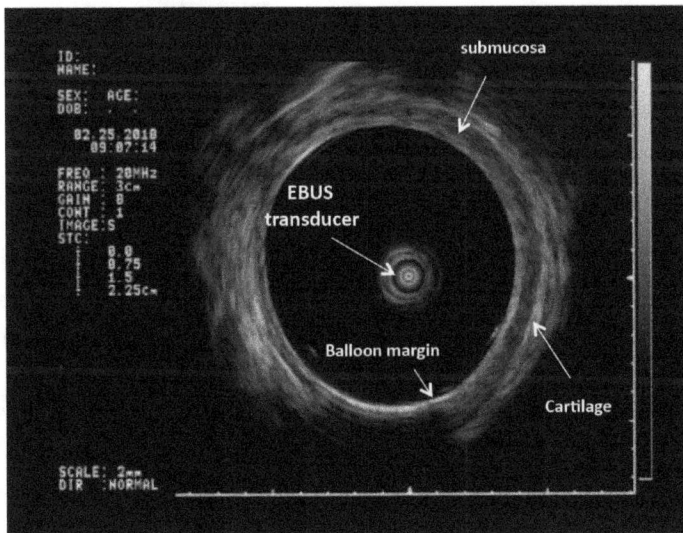

Fig. 4. Endobronchial ultrasound image of normal tracheal wall using a 20-MHz radial probe with axial mechanical scanning. The submucosa and the cartilaginous layers are clearly identified overlying the water filled balloon used for index matching to resolve airway wall-transducer interface.

as for determining the depth of endobronchial tumor invasion.[32,33] Photodynamic therapy (PDT) is less effective in achieving a complete response when tumor extends beyond the cartilaginous layer. More recently, it was shown that in 28% of patients referred for presumed CIS or early cancer, EBUS established disease extent, which would have made endoscopic curative treatment impossible.[34]

EBUS achieves deep penetration of the airway tissue, but the spatial resolution is insufficient for clear demarcation of the microstructural profile and morphologic changes.[35] In addition, due to small differences of acoustic impedance between the healthy and diseased tissues, the imaging contrast offered by EBUS is inherently low. At this time it is possible to perform real time intra-operative endobronchial ultrasonography, but resolutions are typically no higher than 100 μm and physical contact with the tissue or an index matching medium is required to obtain images.

Confocal microscopy

Confocal endoscopy is based on confocal microscopy (CFM) principles, which provide images of a thin section within a biological sample, but in confocal endscopy the microscope's objective is replaced by a flexible fiberoptic miniprobe.[36] In its fluorescence mode, fibered confocal fluorescence microscopy (FCFM), the technique makes it possible to obtain high-resolution images from endogenous or exogenous tissue fluorophores, through a fiberoptic probe that can be introduced into the working channel of a flexible bronchoscope. CFM offers spatial resolution down to the submicron range and depth of penetration of approximately 0.5 mm. Since cellular changes in CIS and dysplasia occur within less than 1 mm in depth, CFM may provide sufficient optical information to distinguish benign from malignant lesions. However, cells do not emit strong autofluorescence. Although the basement membrane and upper submucosa can be imaged with good quality, with the current commercially available technology, the epithelial cells are not visible.[36] Thus inherent tissue and cellular optical contrast remains an issue for FCFM based airway cancer diagnostics. In addition, because contact with the bronchial surface is still required, the fragile epithelium can be scrapped off during the imaging procedure. Motion artifacts due to cardiac pulsation and respiratory movements can also lead to suboptimal imaging of cellular details. Some of these systems are already commercialized and marketed.[36] One system, Cellvizio (Mauna Kea Technologies, Cambridge, MA) is the only currently approved US FDA confocal endoscopic microscopy system for airway use (as of 2010). It uses a 1.4 mm diameter probe for pulmonary applications. The field of view is 600×500 μm, with lateral resolution of 10 μm and a scanning depth of up to 50 μm. Tissue autofluorescence is created by applying a 488 nm excitation light and image acquisition is 12 frames/second.

In vivo applications of FCFM for the airway have recently been studied in patients who underwent bronchoscopy to detect preinvasive lesions.[36] WLB and AF bronchoscopy were performed initially and followed by CFM using a 1.4 mm direct contact probe. AF microstructure alterations were observed in the majority of metaplastic and dysplastic samples and all CIS and invasive cancers. Disorganization of the fibred network was seen indicating basement membrane disruption. Further studies are obviously needed to conform the sensitivity of FCFM for discerning normal from benign and malignant pathologic tissue. Furthermore, since the depth of penetration and field of view is so limited, this technique will likely be appropriate in "multimodality" combination with technologies able to visualize the cartilage and its potential invasion and reveal alterations in the airway wall microstructure such as EBUS and OCT.

Optical Coherence Tomography

Role in early airway cancer detection

The search for a fast acquisition, high spatial resolution, non-invasive technique for endoscopic imaging of *in vivo* tissue structure and function has also led to the development of optical coherence tomography (OCT) systems. Ultrahigh resolution OCT is capable of up to 2 orders of magnitude greater resolution (1–2 μm) than ultrasound (spatial resolution ~50–200 μm) at depths of up to 2 mm in optically scattering tissues (Table 1). The depth range of OCT is sufficient to penetrate through the upper layers of the airway, where many endobronchial carcinomas originate or spread, and is roughly equivalent to the depth that standard endobronchial forceps can sample.[37,38] OCT can employ non-contact probes if desired, avoiding alteration in tissue, while maintaining spatial resolution. The procedure has been reported to add ~5–10 min to a standard bronchoscopic procedure under local anesthesia and moderate sedation. The fiberoptic probes could be potentially miniaturized to enable imaging of airways down to the terminal bronchiole beyond the range of a standard bronchoscope. While definitive diagnosis of cancer is not yet obtainable by OCT, images obtained show potentially characteristic architectural changes in malignancy.[39,40]

OCT definition results from optical refractive index boundaries across the different layers of the airway wall that may be explained in some areas by the presence of a higher nuclear density within structures, such as the epithelium, resulting in enhanced reflectivity signals compared with adjacent surrounding tissues. The extracellular matrix of cartilage, for example, decreases scattering of incident light and thus reflects as a dark region on the OCT image display. (Fig. 5).[40] Because of its maximum penetration depth of 2–3 mm, OCT, however, might not visualize the entire structure of adult human airway cartilage especially in the more proximal airways.

Role in precise determinations of tumor margins

In addition to its potential role in early lung cancer detection, OCT may have a role in identification of tumor boundaries in the mucosa and submucosa of airways. Determining the exact extent of the lesion requires precise delineation of the proximal and distal boundaries of airway involvement. This is an important step in determining operability, extent of resection, and resection margins. Accurate clinical diagnosis of endobronchial cancer peripheral margins is crucial to choosing the correct surgical or medical

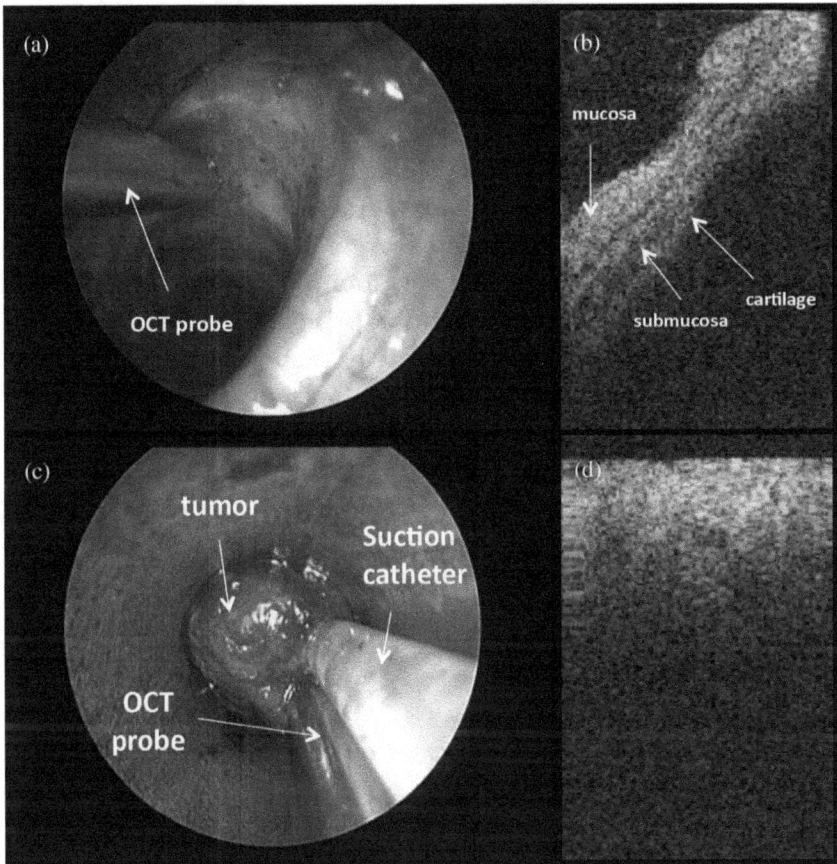

Fig. 5. (a) OCT probe overlies the left mainstem bronchial wall at the level of the cartilaginous ring. (b) The two dimensional time domain OCT image shows the high backscattering layer of the mucosa, overlying the lower backscattering layer of the submucosa. The most inner portion of the cartilage is visualized as a dark region, likely explained by decreased scattering of light in the extracellular matrix of the cartilage. (c) The frontal scanning two dimensional time domain OCT probe and the suction catheter are seen inside the rigid bronchoscope overlying a large endobronchial tumor with hemorrhagic surface. (d) The corresponding OCT image of the tumor shows a bland image with a lack of normal layered microstructures of the airway wall. The hemorrhagic tumor, however, limits image capability because of the absorption of light by hemoglobin and scattering properties of the red blood cells.

therapeutic regimen. Since recurrence at the site of resection must be avoided, preoperative and intraoperative methods for determining the tumor boundaries would be very useful. This is particularly significant when a thoracic resection procedure (thoracotomy or thoracoscopy with lobectomy, sleeve resection, or pneumonectomy) is being considered with curative intent, and ensuring adequacy of staging and surgical margins around a suspected primary malignant lesion is essential. Identification of unsuspected micro-invasion, submucosal spread not visible on the surface, or of small satellite lesions by OCT and AF bronchoscopy could theoretically improve accuracy of biopsy tumor margin assessment, leading to more effective treatment. The role of OCT in such evaluations remains to be determined. However, although OCT may theoretically be very helpful if benign versus malignant airway tissue can be readily distinguished, utility may still be limited for detecting tumor margins due to the relatively shallow depth of penetration, since deeper level tumor invasion cannot be assessed with current OCT technology.[18]

Role in guiding bronchoscopic treatment interventions

Thus far, OCT imaging has been applied for the diagnosis of airway lesions that would often not be evident on gross examination. In the future, by precisely assessing the extent of abnormalities not identified by WLB, one could potentially guide more optimal treatment such as PDT for non-surgical patients with early airway lung cancer through improved site selection, and guidance of diffuser selection. Furthermore, the use of targeted laser energy to induce localized zones of thermal coagulation and necrosis has been investigated *ex vivo* as an intervention for a broad spectrum of disease processes and for *in vivo* in murine tumor studies.[15,41] However, prior studies and current clinical practice have applied these therapies without high resolution imaging guidance, resulting in highly variable outcomes that include both frequent recurrence due to insufficient treatment and complications associated with overly aggressive treatment.

Because OCT is capable of visualizing architectural features of the airway wall, it is well-suited to define therapeutic target volumes *in situ*. A compact, high-resolution, high-speed imaging technique capable of monitoring tissue coagulation, cutting, and ablation intra-operatively at a localized site could allow image guided surgical laser procedures to be monitored for controlled therapy.[15] This may result in subsequent reduction in iatrogenic collateral airway wall injury which is well described in experimental studies.[42] Because OCT images are based on optical backscattering properties of tissue, changes in tissue optical properties during surgical laser ablation should be detectable using this technique. The tissue response to laser exposure is highly dependent on the wavelength of the incident light and the tissue absorption coefficient. In this context, OCT has been proposed to provide feedback that could be used to terminate laser ablation prior to membrane rupture and tissue ejection, demonstrating the ability to monitor tissue response during laser ablation.[15] The ability of OCT to guide laser ablation and image the changes suggests a role in image-guided surgical procedures, such as the bronchoscopic ablation of airway cancer. Further studies are necessary to validate this concept.

In addition to high resolution, OCT is an attractive tool for guiding endoscopic treatment of airway lesions for several other reasons. First, OCT imaging can be fiber based, allowing it to be readily integrated with bronchoscopes. Second, unlike ultrasound, OCT imaging is non-contact and can be performed through air. Third, OCT systems are usually compact and portable, an important consideration for the operating suites. Fourth, OCT imaging can be performed at very high speeds, allowing image data to be acquired rapidly over relatively larger regions of tissue than other high resolution technologies, an important point when patient is undergoing procedures under general anesthesia.

Experimental and clinical OCT studies

The initial reports of the role of OCT in detecting airway preneoplastic and neoplastic lesions involved *ex vivo* studies of normal and abnormal tracheal, bronchial, and bronchiolar anatomy in excised tissue specimens.[39,43,44] Close similarity between OCT imaging and histologic stained excised specimens has been reported for various benign and malignant airway pathology.[45–48] It is noteworthy, however, that OCT-histology correlation is not perfect, a finding partially explained by co-registration issues, and by morphological distortions during histological processing of tissues.[49] Therefore, correlations of OCT images acquired *in vivo* with histopathology from the same site may require tissue and disease-specific correction factors to account for shrinkage and deformations.[50] However, the ability of OCT to differentiate epithelial, subepithelial, glandular, cartilaginous, and muscular structures within the trachea of humans and animal models has been demonstrated and showed reasonable contrast in tissue layer structuring. The initial reports used superluminescent diode, time-domain systems with reported delivered resolutions of 15–30 μm in the axial and lateral directions.

Pitris *et al.* were among the first to evaluate OCT for human airway tissue *in vitro*.[51] In their study, samples from the epiglottis to the secondary bronchi were imaged and compared with histopathology. They verified the ability of OCT to delineate relevant structures such as the epithelium, mucosa, cartilage and glands.

Whiteman *et al.* examined freshly excised lung specimens and showed the ability to differentiate tissue structural characteristics of normal and neoplastic airway using a 10 μm resolution time domain OCT bench-top prototype. However, they were unable, to define individual cellular profiles. As the multilayered anatomy of the healthy airway is disrupted by structural remodeling consequent on development of inflammation and/or neoplasia, OCT tomograms may reflect "bland" images which ignore the normal microstructural boundaries. Inflammation is characterized by a loss of depth beyond the epithelial and lamina propria layers on the OCT image. This partly results from the presence of infiltrating inflammatory cells which have a high nuclear density, thereby modulating the refractive index of the imaged bulk tissue.[40] One limitation in practice however, is that enhanced vascular supply within the inflamed region contributes to reduced imaging capability, as whole blood induces strong attenuation of light. This originates from the absorption of light by hemoglobin and the scattering properties of red blood cells.[52]

In the first clinical report of the endobronchial OCT for lung cancer, layers of the bronchial wall were distinctly observed in the normal bronchus on the OCT images, as opposed to bronchial tumors which lacked a layered structure.[39] At present, most published studies do not include patients with other benign airway pathologies, thus the specificity of characteristic changes seen with airway malignancy is not yet determined.

Studies were developed to determine if OCT could characterize preneoplastic changes in the bronchial epithelium identified by AF bronchoscopy. CIS and invasive carcinoma were able to be distinguished from normal bronchial epithelium. Quantitative measurement of the epithelial thickness showed that invasive carcinoma was significantly different than carcinoma *in situ* and dysplasia was significantly different than metaplasia or hyperplasia. Furthermore, the basement membrane was disrupted or disappeared with invasive carcinoma.[53] Sequential biopsies of the same sites in volunteer smokers with bronchial dysplasia showed a high regression rate at the end of six months in those who were in the placebo arm of the chemo-prevention trial. Since there is still considerable uncertainty about the natural history of these lesions, a non-biopsy method such as OCT might help to clarify the natural history of these lesions and the effect of chemo-preventive intervention or endobronchial therapies.[53]

OCT systems

Commercial as well as prototype OCT systems were used in the initial flexible fiber-optic probe-based *in vivo* reports of airway anatomy in animal models and initial patient studies.[39,45] The commercial systems utilized rotational probes with images obtained at 90° angles from the probe tip. Imalux describes their commercial system probe diameter as 2.7 mm OD using a SLD source with a 10 μm reported resolution that images end-on. Other probes reported for *in vivo* airway investigations were generally on the order of 1.5 mm outer diameter or smaller in order to fit into the working channel of standard bronchoscopes. Transverse (longitudinal imaging) flexible fiberoptic scanning probes were also used to obtain images within trachea and bronchi.[18] There are some theoretical advantages to imaging with transverse scanning probes compared to rotational probes for some applications in large diameter airways, particularly if a combination of larger area imaging, high-resolution, and focus are important. These probes can be kept in contact with the tracheal and bronchial wall surface if desired (helping assure uniform focal distance), and long transverse sweep distances can be obtained with user controlled lateral sampling intervals between axial scans in order to screen large areas. In addition, anatomical image recognition may be somewhat simpler for most pulmonary clinicians in the transverse direction.

Current OCT systems are limited in their ability to define individual cellular profiles. Future developments will need to include faster data acquisition and processing for *in vivo* real-time recording and enhancement of resolution by use of broad-band light sources, and might include OCT-multiphoton microscopy, and/or polarization-sensitive OCT as methods for improving inherent optical contrast definition and obtaining functional information from tissues. Such developments will need to be balanced against constraints placed by the particular *in vivo* milieu in which OCT is applied and the complex tissue profile. Thus, whereas improvements in resolution <2 μm have already been achieved, it may be difficult to enhance imaging depth beyond 2 to 3 mm, again suggesting that future of airway cancer imaging will incorporate multimodal approaches.[54]

In general, motion artifact is not a significant problem in two-dimensional airway OCT acquisition and imaging, particularly if the imaging is done with the probe in a contact mode. The respiratory rate is generally between 5 and 30 breaths per minute in patients undergoing airway examinations. Heart rates vary between approximately 50–150 beats per minute. Therefore, at acquisition rates of 30 frames per second that are readily obtainable using spectral-domain

OCT, result in little or no significant motion artifact is seen at 20 μm resolution. However, if three-dimensional imaging over a reasonably large area is going to be considered, more rapid acquisition methods, or methods to stabilize the tissue region of interest will need to be considered. In addition to contact mode methods, other stabilization methods might be considered for pulmonary and airway OCT three-dimensional imaging in the future.[18]

Future commercial OCT systems might be further optimized for airway applications by miniaturizing the probes so that they can be inserted into standard 2.0 mm and 2.8 mm working channels of flexible bronchoscopes. Such small probes are readily producible with current technology but do add limitations in lateral resolution. In addition, longer sweep distances, broader wavelength laser sources with higher resolution capabilities, compact three-dimensional scanning and integrated rapid Fourier-domain technologies should help improve the sensitivity and specificity of real time OCT airway cancer imaging.

Spectral/Fourier domain OCT

The detection rate in Fourier-domain OCT is enhanced because the receiver registers reflected light from all depth-points in the sample simultaneously over the duration of one complete axial profile. Ultrafast frequency domain OCT has been developed and is limited primarily by the speed at which the actuators need to move the tracking lasers used to scan tissue and photon collection rates.[55] The challenges however, are to access the tissues, acquire data quickly, process the data in real time.[55] The ability to deliver this technology to the region of interest, therefore, is critical for airway cancer diagnostics. This includes development of specialized probes, and most importantly the possibility to miniaturize the technology.

Three-dimensional imaging allows organized image acquisition from volume-based regions. This information is potentially invaluable for three-dimensional airway reconstructions, reconstructions of vascular and functional tissue information, and more thorough and complete investigations of tissue alterations such as malignant transformations. Research groups have reported designs for three-dimensional OCT rigid endoscopic imaging of the airways.[56] Proposed system designs have included: fiber-bundle based designs and rigid GRIN lens probes. Some of these systems allow all movable parts to be placed external (proximal) to the viewing portion of the probe for three-dimensional scanning capabilities, and the potential for dynamic focusing. Similar design concepts could be applied for use in thoracoscopic OCT and general rigid endoscopic OCT probes.

The benefits of Fourier-domain OCT may find utility in screening and surveillance for early airway neoplastic changes. In these applications, the rationale for high-speed acquisition (100 frames/second) is to enable wider-field imaging of large luminal surface areas in the exploration for early stage, focal disease.

Optical frequency domain imaging (OFDI), also known as swept source OCT, is another method that uses a wavelength-swept light source to probe the amplitude and phase of back scattering light from tissue.[60] OFDI uses a spectral domain variation of the optical technology used in OCT with a wavelength swept light source and a stationary reference arm. OFDI offers several advantages, such as higher sensitivity and imaging speed, over conventional time-domain techniques. In combination with a balloon-centering optical catheter, OFDI has been demonstrated to comprehensively image the entire distal esophagus in a time (< 2 minutes, 50 μm pitch) that is acceptable for an endoscopic procedure,[61] and could be readily applied to airway cancer detection.

Emerging Technologies with Potential Applications for Airway Cancer Detection in Conjunction with OCT

There are a number of emerging technologies that may add additional complementary information to that obtained from OCT and become part of multimodality imaging approaches to airway cancer detection. We briefly present them here and speculate on how they may integrate with information obtained by OCT for airway cancer detection in the future.

Raman spectroscopy imaging

Raman spectroscopy is based on Raman scattering of light from molecules. Inelastic scattering is a fundamental scattering process in which the kinetic energy of an incident photon is not conserved. In this scattering process, the energy of the incident particle is lost or gained. Because the wavelength of Raman scattered light depends on molecular composition, Raman spectra provide highly useful information about molecular composition. The technique relies on the scattering of monochromatic light, usually delivered by a near-infrared (NIR) laser having a wavelength of 750–1400 nm.[62] The difference between the incident and scattered frequencies corresponds to the vibrational modes of molecules participating in the interaction. Depth of penetration can vary between 0.5–1 mm and the acquisition time between 1 and 5 seconds depending on the system used.[62,63]

Studies show that specific features of tissue Raman spectra can be related to the molecular and structural changes associated with neoplastic transformations.[64,65] A sensitivity and specificity of 82% and 92%, respectively, for differentiating between precancerous and benign cervical tissue *in vitro*, has been reported.[66] *In vivo* NIR Raman measurements have also been reported for the cervix, colon, esophagus, skin and recently for the bronchus.[62] Thus, distinctive Raman features and intensity differences for tumor versus normal bronchial tissue can reflect molecular and cellular changes associated with malignant transformation. The sensitivity of cancer prediction is reported to be as high as 91%, with 97% specificity.[63] These experimental *in vitro* studies show that there appear to be specific differences for malignant tumor versus normal tissue, supporting a potential role for NIR Raman spectroscopy in lung cancer diagnosis based on probing changes at molecular level. To date, no confirmatory *in vivo* studies have been published, and again, such studies would also have to include patients with other benign pathologies in addition to malignant and normal cases in order to determine sensitivity and specificity.

The main drawback of the Raman spectroscopy is the small sampling area of the probe and penetration depth. Similar to other high-resolution technologies, only small volumes of interrogation are obtainable. It is impossible to scan the whole bronchial tree with the Raman probe. Therefore, if ever used for early lung cancer detection, Raman spectroscopy will have to be applied in conjunction with other more sensitive modalities such as, OCT, AF bronchoscopy, and NBI. This is a contact mode technology and blood, mucus, and even tissue's autofluorescence may interfere with this process. As technology of the Raman spectroscopy advances, catheter and detector advancements may facilitate its role in bronchoscopic detection of early lung cancer.

Photoacoustics

Commercially available high-resolution three-dimensional optical imaging modalities including CFM and OCT have impacted biomedicine but such tools cannot penetrate biological tissue deeper than 2–3 mm. Photoacoustic tomography (PAT) or photoacoustic microscopy (PAM) which combines strong optical contrast and high ultrasonic resolution in a single modality, has overcome this depth limitation and achieved "superdepth" high-resolution optical imaging.[67] PAT is cross-sectional or three-dimensional imaging based on the photoacoustic effect (also called optoacoustic or thermoacoustic). The tissue is usually irradiated by a short pulsed laser beam to produce thermal and acoustic impulse responses. Locally absorbed light is converted into heat, which is further converted to a pressure rise via thermoelastic expansion of the tissue. The initial pressure rise, determined by the local optical energy deposition and other thermal and mechanical properties, propagates in the tissue as an ultrasonic wave, which is referred to as a photoacoustic wave, which is then detected by ultrasonic transducers.[67] By using multiwavelength measurement, one can simultaneously quantify concentrations of multiple chromophores of different colors, such as oxygenated and deoxygenated hemoglobin molecules in red blood cells. Such quantification of hemoglobin can provide functional imaging of the concentration and oxygen saturation of hemoglobin. Both parameters are related to hallmarks of cancer: the concentration of hemoglobin correlates with angiogenesis, whereas the oxygen saturation of hemoglobin correlates with hypermetabolism.[67,68] PAT may help fill the void of penetration and reached super depth optical imaging. Because it does complement the other high resolution optical technologies in both penetration and contrast, PAT may become a useful tool in biomedical research and potentially in clinical practice.

Practical Development Needs for Clinical Airway OCT Based Technology

There are several OCT technological development needs that, if fulfilled, would markedly improve the utility of OCT for airway diagnostics. Many of these concepts are being actively pursued at this time by research groups for pulmonary and other general applications. The following are several of the major areas for further development.

Improved resolution

Practical methods for obtaining 1–2 μm resolution and improved optical contrast at depths of millimeters and more are needed for endobronchial diagnostics in order to resolve nuclear components to more reliably distinguish benign from malignant cells. While tissue structural properties may provide some information on likelihood of malignancy, more definitive differentiation requires visualization of nuclear structure, or cancer related optical signals. It remains to be determined whether endogenous optical contrast will be high enough to visualize nuclei *in vivo* when ultrahigh resolution OCT of the lungs becomes readily available. Affordable, reliable, broadband laser light sources and rapid swept-source laser technologies that can be used in a clinical setting are rapidly advancing at this time. However, in addition to the optical sources, flexible and rigid probes of relatively narrow diameters capable of delivering these high resolutions are also needed. Such probes will likely require development of dynamic focusing methods in order to preserve the high resolutions throughout the depths of the imaging axial scans.

Concurrent OCT imaging and biopsy tools

A major limitation of current OCT probes applicable to airway diagnostics is the inability to obtain a biopsy concurrent with real-time OCT imaging. This is particularly problematic in flexible fiberoptic probe based systems. Analogous problems are encountered with the use of radial probe EBUS. The ability to visualize structures under OCT and obtain immediate biopsy without removal of the probe would be of great value. Such systems will need to be designed for rigid endoscopic probes, flexible fiberoptic probes, and eventually for needle based probe systems.

Optical biopsy

Some innovators in the development OCT have proposed the concept of "optical biopsy" in which OCT high resolution imaging may replace current traditional tissue biopsy methods.[38] While this is a valuable long-term research target, even ultrahigh resolution OCT imaging is unlikely to replace tissue biopsy for pulmonary diagnostics in the near future. While biopsy specimen examination for histological appearance may soon be rivaled by ultrahigh resolution optical techniques, a range of other information obtained from tissue biopsy may still be unobtainable until considerable scientific advances have been made. For example, special immunologic staining for cellular antigens to characterize tissue origin, define histologic subtypes, as well as chemosentitivity testing, DNA phenotyping, or infectious agent evaluation currently require excised tissue for staining and/or culture and sensitivity analysis. While it is conceivable that *in vivo* molecular signals with analogous capabilities may someday become available to obviate these needs, the broad range of advances necessary will undoubtedly take many years to develop.[18]

Conclusions

We envision that a major clinical role for OCT in airway cancer will likely include early detection, guiding biopsy sites, bronchoscopic therapies and assessing response to therapeutic interventions in the near-term. Multimodality imaging may use AF, NBI, HMB, EBUS, CFM and OCT in various combinations based on clinical application needs, location, and resolution requirements.[70] In this way OCT may be ideal for early proximal airway malignancy detection when combined with field narrowing AF or NBI modalities. In the longer term, practical peripheral lung parenchymal cancer detection with OCT will require development of small, high resolution needle based systems.

OCT may be of assistance in guiding endobronchial biopsy sites, improving biopsy yield for endobronchial lesions, and reducing complications. OCT alone or in combination with other high resolution optical technologies mentioned in this chapter may be useful in identifying areas of high yield within large mucosal and submucosal tumors of the proximal airways as well as potentially guiding laser or photodynamic treatment. The study of each individual imaging modality is rapidly progressing and is an essential step in the process of multimodality bronchoscopic airway imaging development for airway cancer detection and other applications.

References

1. U.S. Cancer Statistics Working Group. (2010) United States Cancer Statistics: 1999–2006 Incidence and Mortality Web-based Report. Atlanta (GA): Department of Health and Human Services, Centers for Disease Control and Prevention, and National Cancer Institute. Available at: *http://www.cdc.gov/uscs*
2. Jemal A, Siegel R, Ward E *et al.* (2007) Cancer statistics, 2007. *CA Cancer J Clin* **57**: 43–66.
3. Colt HG, Murgu SD. (2010) Interventional bronchoscopy from bench to bedside: New techniques for early lung cancer detection. *Clin Chest Med* **31**: 29–37.

4. Auerbach O, Stout AP, Hammond EC, Garfinkel L. (1961) Changes in bronchial epithelium in relation to cigarette smoking and in relation to lung cancer. *New Engl J Med* **265**: 253–268.

5. Rom WN, Hay JG, Lee TC *et al.* (2000) Molecular and genetic aspects of lung cancer. *Am J Respir Crit Care Med* **161**: 1355–1367.

6. Ikeda N, Hayashi A, Iwasaki K *et al.* (2007) Comprehensive diagnostic bronchoscopy of central type early stage lung cancer. *Lung Cancer* **56**: 295–302.

7. Band PR, Feldstein M, Saccomanno G. (1986) Reversibility of bronchial marked atypia: Implication for chemoprevention. *Cancer Detect Prev* **9**: 157–160.

8. Venmans BJ, van Boxem TJ, Smit EF *et al.* (2000) Outcome of bronchial carcinoma *in situ. Chest* **117**: 1572–1576.

9. Jeremy GP, Banerjee AK, Read CA *et al.* (2007) Surveillance for the detection of early lung cancer in patients with bronchial dysplasia. *Thorax* **62**: 43–50.

10. Doi K. (2005) Current status and future potential of computer-aided diagnosis in medical imaging. *Br J Radiol* **78**: S3–S19.

11. Schaefer-Prokop C, Prokop M. (2002) New imaging techniques in the treatment guidelines for lung cancer. *Eur Respir J Suppl* **35**: 71s–83s.

12. Finkelstein SE, Summers RM, Nguyen DM, Schrump DS. (2004) Virtual bronchoscopy for evaluation of airway disease. *Thorac Surg Clin* **14**: 79–86.

13. Muller NL. (2002) Computed tomography and magnetic resonance imaging: Past, present and future. *Eur Respir J Suppl* **35**: 3s–12s.

14. Murgu SD, Colt HG. (2006) Tracheobronchomalacia and excessive dynamic airway collapse. *Respirology* **11**: 388–406.

15. Boppart S, Herrmann J, Pitris C *et al.* (1999) High-resolution optical coherence tomography-guided laser ablation of surgical tissue. *J Surg Res* **82**: 275–284.

16. Mac Manus MP, Hicks RJ. (2003) PET scanning in lung cancer: Current status and future directions. *Semin Surg Oncol* **21**: 149–155.

17. Koh DM, Burke S, Davies N, Padley SP. (2002) Transthoracic US of the chest: Clinical uses and applications. *Radiographics* **22**: e1.

18. Brenner M, Mahon S, Colt H, Chen Z. (2008) Optical coherence tomography in pulmonary medicine. In: Drexler W, Fujimoto JG (eds.),

Optical Coherence Tomography: Technology and Applications, pp. 1183–1209. Springer, Berlin, Heidelberg.

19. West JB. (ed.) (1985) *Respiratory Physiology — The Essentials,* 3rd ed. Williams & Wilkins, Baltimore.

20. Popovich J Jr, Kvale PA, Eichenhorn MS *et al.* (1982) Diagnostic accuracy of multiple biopsies from flexible fiberoptic bronchoscopy. A comparison of central versus peripheral carcinoma. *Am Rev Respir Dis* **125**: 521–523.

21. Häussinger K, Pichler J, Stanzel F, *et al.* (2000) Autofluorescence bronchoscopy: The D-Light System. In Bolliger CT, Mathur PN (eds.), *Interventional Bronchoscopy. Prog Respir Res* pp. 243–252. Karger, Basel.

22. Gabrecht T, Glanzmann T, Freitag L *et al.* (2007) Optimized autofluorescence bronchoscopy using additional backscattered red light. *J Biomed Opt* **12**: 064016.

23. Furukawa K, Ikeda N, Miura T *et al.* (2003) Is autofluorescence bronchoscopy needed to diagnose early bronchogenic carcinoma? Pro: autofluorescence bronchoscopy. *J Bronchol* **10**: 64–69.

24. Pierard P, Vermylen P, Bosschaerts T *et al.* (2000) Synchronous roentgenographically occult lung carcinoma in patients with resectable primary lung cancer. *Chest* **117**: 779–785.

25. Weigel TL, Yousem S, Dacic S *et al.* (2000) Fluorescence bronchoscopic surveillance after curative surgical resection for non-small-cell lung cancer. *Ann Surg Oncol* **7**: 176–180.

26. Bota S, Auliac JB, Paris C, *et al.* (2001) Follow-up of bronchial precancerous lesions and carcinoma in situ using fluorescence endoscopy. *Am J Respir Crit Care Med* **164**: 1688–1693.

27. Shibuya K, Hoshino H, Chiyo M, *et al.* (2002) Subepithelial vascular patterns in bronchial dysplasias using a high magnification bronchovideoscope. *Thorax* **57**: 902–907.

28. Vincent BD, Fraig M, Silvestri GA. (2007) A pilot study of narrow-band imaging compared to white light bronchoscopy for evaluation of normal airways and premalignant and malignant airways disease. *Chest* **131**: 1794–1799.

29. Herth FJ, Eberhardt R, Anantham D *et al.* (2009) Narrow-band imaging bronchoscopy increases the specificity of bronchoscopic early lung cancer detection. *J Thorac Oncol* **4**: 1060–1065.

30. Bamber JC, Tristam M. (1988) Diagnostic ultrasound. In: S Webb (ed), *The Physics of Medical Imaging*, pp. 319–388. Adam Hilger Bristol, Philadelphia.

31. Sutedja TG, Codrington H, Risse EK *et al.* (2001) Autofluorescence bronchoscopy improves staging of radiographically occult lung cancer and has an impact on therapeutic strategy. *Chest* **120**: 1327–1332.

32. Herth F, Ernst A, Schulz M, Becker H. (2003) Endobronchial ultrasound reliably differentiates between airway infiltration and compression by tumor. *Chest* **123**: 458–462.

33. Kurimoto N, Murayama M, Yoshioka S, Nishisaka T. (1999) Assessment of usefulness of endobronchial ultrasonography in determination of depth of tracheobronchial tumor invasion. *Chest* **115**: 1500–1506.

34. Miyazu Y, Miyazawa T, Kurimoto N *et al.* (2002) Endobronchial ultrasonography in the assessment of centrally located early-stage lung cancer before photodynamic therapy. *Am J Respir Crit Care Med* **165**: 832–837.

35. Goldberg BB, Liu JB, Merton DA *et al.* (1993) Sonographically guided laparoscopy and mediastinoscopy using miniature catheter-based transducers. *J Ultrasound Med* **12**: 49–54.

36. Thiberville L, Moreno-Swirc S, Vercauteren T *et al.* (2007) *In-vivo* imaging of the bronchial wall microstructure using fibered confocal fluorescence microscopy. *Am J Respir Crit Care Med* **175**: 22–31.

37. Drexler W, Morgner U, Kärtner FX *et al.* (1999) *In-vivo* ultrahigh-resolution optical coherence tomography. *Optics Lett* **24**: 1221–1223.

38. Fujimoto JG, Brezinski ME, Tearney GJ *et al.* (1995) Optical biopsy and imaging using optical coherence tomography. *Nat Med* **1**: 970–972.

39. Tsuboi M, Hayashi A, Ikeda N *et al.* (2005) Optical coherence tomography in the diagnosis of bronchial lesions. *Lung Cancer* **49**: 387–394.

40. Whiteman SC, Yang Y, Gey van Pittius D *et al.* (2006) Optical coherence tomography: Real-time imaging of bronchial airways microstructure and detection of inflammatory/neoplastic morphologic changes. *Clin Cancer Res* **12**: 813–818.

41. Chen Z, Milner TE, Wang X, Srinivas S, Nelson JS. (1998) Optical Doppler tomography: Imaging *in-vivo* blood flow dynamics following pharmacological intervention and photodynamic therapy. *Photochem Photobiol* **67**: 56–60.

42. Jung Kwon O, Young Suh G, Pyo Chung M, Kim J, Han J, Kim H. (2003) Tracheal stenosis depends on the extent of cartilaginous injury in experimental canine model. *Experimental Lung Research* **29**: 329–338.

43. Yang Y, Whiteman S, van Pittius DG, He Y, Wang RK, Spiteri MA. (2004) Use of optical coherence tomography in delineating airways microstructure: Comparison of OCT images to histopathological sections. *Phys Med Biol* **49**: 1247–1255.

44. Gossage KW, Tkaczyk TS, Rodriguez JJ, Barton JK. (2003) Texture analysis of optical coherence tomography images: Feasibility for tissue classification. *J Biomed Opt* **8**: 570–575.

45. Hanna N, Saltzman D, Mukai D *et al.* (2005) Two-dimensional and three-dimensional optical coherence tomographic imaging of the airway, lung, and pleura. *J Thorac Cardiovasc Surg* **129**: 615–622.

46. Jung W, Zhang J, Mina-Araghi R *et al.* (2004) Feasibility study of normal and septic tracheal imaging using optical coherence tomography. *Lasers Surg Med* **35**: 121–127.

47. Colt H, Murgu SD, Ahn YC, Brenner M. (2009) Multimodality bronchoscopic imaging of tracheopathica osteochondroplastica. *J Biomed Opt* **14**: 034035.

48. Colt HG, Murgu SD, Jung B, Ahn YC, Brenner M. (2010) Multimodality bronchoscopic imaging of recurrent respiratory papillomatosis. *Laryngoscope* **120**: 468–472.

49. Bloom G, Friberg U. (1956) Shrinkage during fixation and embedding of histological specimens. *Acta Morphol Neerl Scand* **1**: 12–20.

50. Hsiung PL, Nambiar PR, Fujimoto JG. (2005) Effect of tissue preservation on imaging using ultrahigh resolution optical coherence tomography. *J Biomed Opt* **10**: 064033.

51. Pitris C, Brezinski ME, Bouma BE, Tearney GJ, Southern JF, Fujimoto JG. (1998) High resolution imaging of the upper respiratory tract with optical coherence tomography: A feasibility study. *Am J Respir Crit Care Med* **157**: 1640–1644.

52. Villard JW, Feldman MD, Kim J, Milner TE, Freeman GL. (2002) Use of a blood substitute to determine instantaneous murine right ventricular thickening with optical coherence tomography. *Circulation* **105**: 1843–1849.

53. Lam S, Standish B, Baldwin C *et al.* (2008) *In-vivo* optical coherence tomography imaging of preinvasive bronchial lesions. *Clin Cancer Res* **14**: 2006–2011.

54. Lim H, Jiang Y, Wang Y, Huang YC, Chen Z, Wise FW. (2005) Ultrahigh-resolution optical coherence tomography with a fiber laser source at 1 microm. *Opt Lett* **30**: 1171–1173.

55. Bouma BE, Yun SH, Vakoc BJ, Suter MJ, Tearney GJ. (2009) Fourier-domain optical coherence tomography: recent advances toward clinical utility. *Curr Opin Biotechnol* **20**: 111–118.

56. Xie T, Mukai D, Guo S, Brenner M, Chen Z. (2005) Fiber-optic-bundle-based optical coherence tomography. *Opt Lett* **30**: 1803–1805.

57. Huber R, Adler DC, Fujimoto JG. (2006) Buffered Fourier domain mode locking: Unidirectional swept laser sources for optical coherence tomography imaging at 370,000 lines/s. *Opt Lett* **31**: 2975–2977.

58. Potsaid B, Gorczynska I, Srinivasan VJ *et al.* (2008) Ultrahigh speed spectral/Fourier domain OCT ophthalmic imaging at 70,000 to 312,500 axial scans per second. *Opt Express* **16**: 15149–15169.

59. Oh WY, Yun SH, Tearney GJ, Bouma BE. (2005) 115 kHz tuning repetition rate ultrahigh-speed wavelength-swept semiconductor laser. *Opt Lett* **30**: 3159–3161.

60. Yun SH. (2006) Optical coherence tomography using rapidly swept lasers. *Conf Proc IEEE Eng Med Biol Soc* **1**: 125–128.

61. Suter MJ, Jillella PA, Vakoc BJ *et al.* (2010) Image-guided biopsy in the esophagus through comprehensive optical frequency domain imaging and laser marking: A study in living swine. *Gastrointest Endosc* **71**: 346–353.

62. Huang Z, McWilliams A, Lui H, McLean DI, Lam S, Zeng H. (2003) Near-infrared Raman spectroscopy for optical diagnosis of lung cancer. *Int J Cancer* **107**: 1047–1052.

63. Yamazaki H, Kaminaka S, Kohda E, Mukai M, Hamaguchi HO. (2003) The diagnosis of lung cancer using 1064-nm excited near-infrared multichannel Raman spectroscopy. *Radiat Med* **21**: 1–6.

64. Kawabata T, Mizuno T, Okazaki S *et al.* (2008) Optical diagnosis of gastric cancer using near-infrared multichannel Raman spectroscopy with a 1064-nm excitation wavelength. *J Gastroenterol* **43**: 283–290.

65. Teh SK, Zheng W, Ho KY, Teh M, Yeoh KG, Huang Z. (2008) Diagnostic potential of near-infrared Raman spectroscopy in the stomach: Differentiating dysplasia from normal tissue. *Br J Cancer* **98**: 457–465.

66. Mahadevan-Jansen A, Mitchell MF, Ramanujam N *et al.* (1998) Near-infrared Raman spectroscopy for *in vitro* detection of cervical precancers. *Photochem Photobiol* **68**: 123–132.

67. Wang LV. (2008) Prospects of photoacoustic tomography. *Med Phys* **35**: 5758–5767.

68. Li C, Wang LV. (2009) Photoacoustic tomography and sensing in biomedicine. *Phys Med Biol* **54**: R59–97.

69. Karpiouk AB, Wang B, Emelianov SY. (2010) Development of a catheter for combined intravascular ultrasound and photoacoustic imaging. *Rev Sci Instrum* **81**: 014901.

70. DaCosta RS, Wilson BC, Marcon NE. (2005) Optical techniques for the endoscopic detection of dysplastic colonic lesions. *Curr Opin Gastroenterol* **21**: 70–79.

Chapter 7

Diffuse Optical Spectroscopy and Imaging in Breast Cancer

Albert E. Cerussi[*,†] *and Bruce J. Tromberg*[*,‡]

Introduction

The definitive breast cancer diagnosis is based upon pathology. The pathological origins of contrast, both structural and cytological, manifest as alterations in tissue optical properties, mainly in scattering. Key physiological processes typical of malignant transformation that are accessed via biomarkers, such as angiogenesis, increased proliferation, metabolic transformation and inflammation also alter tissue optical properties, mainly in absorption.

Microscopic functional and structural alterations to cells, vessels and matrix can be probed using non-invasive macroscopic optical measurements. Although the specificity of pathology and biomarker measurements is diminished, diffuse optical approaches have shown that absorption and scattering contrast mechanisms are preserved across spatial scales. In this contribution we review how non-invasive diffuse optical measurements of tissue optical properties, namely absorption (μ_a) and "reduced" scattering (μ_s') parameters are addressing clinical problems in breast cancer.

* Beckman Laser Institute, University of California, Irvine, 1002 Health Sciences Road, East, Irvine, CA 92612, USA
† E-mail: acerussi@uci.edu
‡ E-mail: bjtrombe@uci.edu

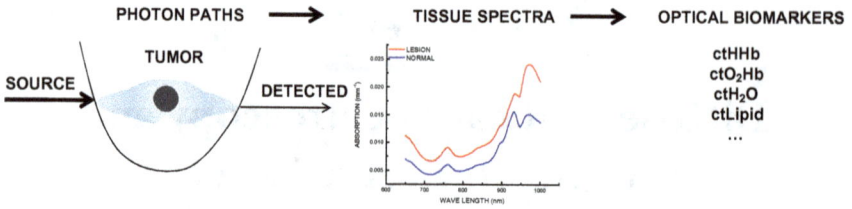

Fig. 1. *Left*: A broad range of different photon path lengths are detected, with some photons interacting with the tumor. *Center*: After proper modeling, tissue absorption and scattering spectra are measured using the measured path lengths. *Right*: Optical biomarkers such as deoxy-hemoglobin are calculated from the measured spectra.

Nomenclature

"Diffuse" refers to multiply scattered photons that behave as stochastic particles propagating according to a photon density gradient.[1,2] Reliance on multiply-scattering photons for contrast is quite different from high-spatial resolution modalities such as x-ray mammography. Intense multiple scattering of light by tissues diminishes imaging spatial resolution (5–10 mm) and probes a large tissue volume (Fig. 1).[3] The shaded area in the figure represents a distribution of most probable photon paths that will pass through the tissue and be detected; the spatial broadening is inherent in the light transport. However, this limitation is offset by enormous functional sensitivity. Diffuse optics spatial resolution is closer to nuclear imaging methods such as PET and SPECT, rather than anatomic methods such as mammography and MRI.

The terminology for diffuse optical methods has varied over the years. Early approaches referred to "diaphanography", lightscanning, and Near Infrared Spectroscopy (NIRS). More recent examples include Diffuse Optical Tomography (DOT), Diffuse Optical Imaging (DOI), Diffuse Optical Spectroscopy (DOS), Photon Migration Imaging, and Optical Mammography. The use of "Diffuse Optics" rather than NIRS typically signifies quantitative methods based on photon diffusion models. In this review we employ DOS or DOI to draw parallels with Magnetic Resonance Spectroscopy (MRS) and Imaging (MRI) and emphasize model-based quantitative analysis.

Near infrared spectra inform tissue physiology

Tissue absorption and scattering properties link the physics of light propagation with the biology of tissue structure and function. Near-infrared (NIR) photons (~650–1000 nm) penetrate deeply into tissues (e.g. several centimeters). The NIR absorption spectrum features distinctive contributions from three primary molecules: hemoglobin (Hb), water (H_2O) and lipids.[4–5] Concentrations of primary NIR absorbers and their sub-states are quantified via the standard Beer-Lambert Law commonly employed in any spectrophotometer. The main difference between spectrophotometric measurements performed in a cuvette vs. tissue is the substantially increased photon path length in tissues that results from intense multiple light scattering (Fig. 1).

Representative spectra of bulk breast tissue are provided in Fig. 2. The plots are a statistical average of 58 malignant and normal breast tissues measured *in vivo* using broadband DOS.[4] NIR spectra demonstrate stark physiological differences between tumor and normal

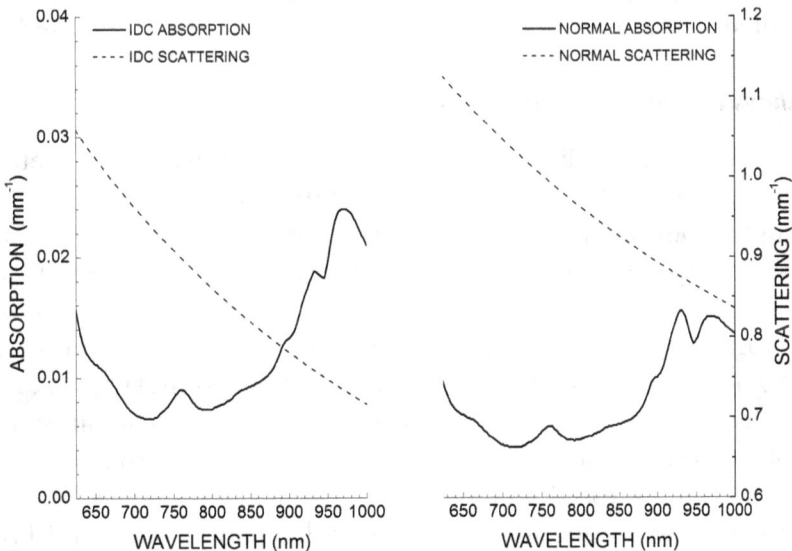

Fig. 2. Average bulk absorption and scattering spectra for infiltrating ductal carcinoma (IDC, left) and normal (right) breast tissues.[4]

tissues. Below 850 nm tumors display increased absorption from hemoglobin relative to normal tissues. These hemodynamic signals originate from the enhanced angiogenesis of malignancy, generally from smaller vessels. Near 980 nm, tumors display increased absorption due to increased water, as well as spectral shifts suggestive of changes in water binding state.[4,6] This signal originates from edema and increased cellularity in malignant tissues. The peak in the normal absorption spectrum at 930 nm is due to bulk lipids, a signal that diminishes as malignant tumors grow. The wavelength dependence of tissue scattering (i.e., scatter power, SP) also differs between tumor and normal tissues, suggesting that malignancy changes the density and size of tissue scattering centers.[7] This signal originates from malignant cell morphology as well as breast tissue collagen density.[8]

Physiological values, and hence spectral values, are highly dependent upon age, hormonal status and other patient physiological parameters.[4,9] Generally, pre-menopausal tissues display higher absorption and scattering than post-menopausal tissues owing to higher metabolic activity. Spectra may change during the menstrual cycle, though the effects are highly variable and not well documented.[10] Tumor optical properties also seem to scale with age.[4]

Exhaustive parameters list for NIR spectra

Table 1 provides a list of the major known NIR tissue parameters, which are separated into categories by the dotted horizontal lines; units for some are provided in parentheses. The "ct" prefix denotes "tissue concentration". The first category represents basic tissue absorbers. Generally every DOS/DOI instrument measures deoxy-hemoglobin (ctHHb) and oxy-hemoglobin (ctO$_2$Hb) concentrations in some capacity, either relative or absolute (micromolar, μM). Two wavelengths are required to measure these hemoglobin states, though accuracy lessens with reduced spectral information content.[11] With increased spectral bandwidth, generally above 900 nm, the concentration of additional absorbers such as water and lipid (% of pure substance) can be calculated.[4,12] Preliminary reports of collagen absorption have also appeared using wavelengths above 1000 nm, but they have not been widely implemented or validated.[13]

Table 1. Summary of NIR contrast parameters for DOS and DOI.

Tissue parameter	Symbol	Description
Deoxy Hemoglobin (μM)	ctHHb	Metabolism/consumption
Oxy Hemoglobin (μM)	ctO$_2$Hb	Metabolism/delivery
Water (% of pure water)	ctH$_2$O	Hydration/edema
Lipid (% of pure lipid)	ctLipid	Bulk lipid signal
Collagen (mg/cm^3)	ctCol	Collagen type 1
Total Hb (μM)	ctTHb	Blood volume
Hemoglobin Saturation (%)	stO$_2$	Oxygenation status
Tissue Optical Index (μM)	TOI	Metabolism
Bound Water	BWI	Macromolecule binding
Specific Tumor Component	STC	Lesion-specific absorption
Scatter Power	SP	Scatter size (cell/matrix)
Scatter Amplitude	A	Scatter density (cell/matrix)

Note: not all systems measure each of these parameters.

Absorption-based indices are also employed in order to condense information and enhance physiological insight (category #2). The most common are total hemoglobin concentration (ctTHb = ctHHb + ctO$_2$Hb) and hemoglobin saturation (stO$_2$ = ctO$_2$Hb/ctTHb × 100%). We have developed a Tissue Optical Index (TOI = ctHHb × ctH$_2$O/ctLipid) that provides high contrast for tumors in breast tissues.[4]

Increased broadband spectral information content has led to the discovery of dispositional shifts in NIR absorbing molecules mainly water and lipids (the third category). Using a spectral referencing technique that normalizes an individual's tissue physiology, absorption spectra unique to lesions have been detected in breast tissues.[14] These unique absorption spectra, termed the "specific tumor component (STC)" act as a breast lesion fingerprint likely originating from alterations in lipid metabolism and water binding state. Further, the STC may discriminate between benign and malignant breast lesions.[15] Water dispositional changes can be characterized using the "bound water index (BWI)." The absorption spectra of water bound and free states are shifted due to variations in the relative contributions of harmonic overtones from fundamental O–H vibrations at 3.05 μm and 2.87 μm. The balance between bound and free water states has been shown to change in malignancy, mainly due to increases in free water.[6]

In the fourth category, parameters are derived from μ_s'. Values are sometimes reported as the "scatter power (SP)" or "scatter amplitude (A)" according to the relationship $\mu_s' = A \times \lambda^{-SP}$.[7] In the macroscopic[4,8] and microscopic[16,17] tissue realms some studies have shown these parameters change in cancer but there is much to be explored. Many NIR instruments do not measure tissue scattering. Further, separation between absorption and scattering is sometimes incomplete as a consequence of instrumentation and modeling limitations.

Brief history of diffuse optics in breast

Although white-light transillumination was introduced into medicine in the early 1800's, it was not until the 1920's that the technique was applied to detect breast lesions. A light source was held underneath the breast in a dark room, and the attenuation by blood created visible shadows from blood vessels and breast lesions. The method was qualitative, coming before mammography. Spectral breast imaging was introduced in the 1980's by restricting the light source to red and NIR bands. This increased penetration depth but reduced visible contrast, plus new electronic detectors were used to enhance sensitivity. Images consisted of the ratios between red and NIR light traveling through the breast. Clinical trials attempted to compare transillumination instruments with mammography as a screening tool.[18–20] Mixed results dissuaded further development.

The limitations of diaphanography were very high false positive and low true positive rates. Many reasons contributed to the failures of these early optical methods to detect and to classify breast lesions: such as study-dependent issues (i.e., poor operator experience, and inadequate statistical analysis) and physical limitations (i.e., inadequate light penetration, poor dynamic range, inadequate spatial resolution, uncorrected geometrical/boundary effects and tissue scattering).

The late 1980's and 1990's witnessed a renewed enthusiasm for optical mammography because of advances in both light-tissue interaction theory and optical technology. Light propagation modeling in tissue is a key aspect of DOS/DOI and not diaphanography. Research has focused upon utilizing radiation transport models,[2, 21] implementing pulsed (i.e., time domain) or intensity modulated light sources (i.e., frequency domain), and model-based methods for quantifying absorption and scattering.[1,22–25] Technological improvements (i.e., laser diodes, solid-state detectors, computer processing) further contributed to the growth of DOS/DOI.

Types of diffuse optical instruments

There are a variety of DOS/DOI research instruments from commercial and academic labs. Only standardized clinical testing will transition these instruments into clinical use. Our list is representative, not exhaustive; we report published instruments that have been used in some sort of clinical test (loosely defined as normal and lesion patients in some capacity). Only endogenous contrast systems are described.

Table 2 lists published DOS/DOI instruments grouped between the dotted lines by imaging geometry: first, handheld reflectance (like ultrasound), second, parallel plate transmission (like mammography), and third, ring/circular (like MRI). The measurement modality is also displayed: Continuous Wave (CW), Frequency Domain (FD), and Time Domain (TD). General descriptions and advantages of these approaches have been documented.[26] The number of sources (s) and detectors (d) are listed, or alternatively the number of source-detector pairs (s/d).

The majority of instruments in the literature are NIR imagers that employ arrays of optical fibers or cameras.[12,27–35] Handheld probes typically have fewer s/d pairs, but offer some advantages over fixed imaging systems, such as region of interest control and high portability. Mild breast compression is typically employed in planar

Table 2. Summary of NIR diffuse optical instruments (see text for legend).

Description	Mode	Imaging	Bandwidth (nm)	Notes
"Puck" Imager[46]	CW	8s × 1d	760/805/850	Portable
P-Scan[37]	CW	8s × 8d	690/830	+Compress/Flow
U. Conn. DOT[92]	FD	12s × 8d + slices	780/830	+Ultrasound
UC Irvine DOS[4]	FD+CW	1s × 1d	650–1000 (1024 λ's)	Portable
ComfortScan[28]	CW	127s + CCD	640	+Compress/Flow
U. Florida DOT[52]	CW	64s × 64d (3D)	10 λ (638-965)	
LIMA/Zeiss[69]	FD	1s × 1d (raster scan)	690/710	
Siemens[51]	FD	1s × 1d (raster scan)	690/750/788/856	
TOBI[36]	FD+CW	>2 × 10^3 pairs	685/810/830	+Tomosynthesis
Penn DOT[27]	FD+CW	~4 × 10^4 pairs (3D)	690/750/786/830/650/905	
SoftScan[29]	TD	80 × 95 mm (1520pts)	760/780/830/850	
Berlin DTOF[33]	TD	~2 × 10^3 scan	670/780/843 or 884	
Milan MAMMOT[12]	TD	1 mm step scan	637, 785, 905, 975	
Phillips DOT[31]	CW	253s/254d	690/730/780/850	
CTLM[30]	CW	84d (scan)	808	Scanning ring
UCL[34]	TD	32s/32d	780/815	
Dartmouth DOT[8]	FD	16s × 15d × 3 rings		+MRI

imaging instruments. Ring geometries typically employ 8, 16 or 32 s/d pairs or sometimes a cup with many s/d pairs. The effects of boundary conditions (i.e., tissue to air interface) are sometimes minimized by the use of index matching fluids.

Most instruments use multiple wavelengths to measure hemoglobin. At the minimum, two wavelengths in the 650–850 nm range are employed, one above and one below the hemoglobin isobestic point at 800 nm. In this region, laser diodes are plentiful and detectors possess good sensitivity. Some instruments have spectral coverage into the 900+ range to measure water and/or lipids. Some instruments also employ dynamic compression to modulate blood flow and thus focus the contrast to regions of high blood vessel density. This approach has been employed in both parallel plate [28,36] and handheld configurations.[37]

DOI inherently has low spatial resolution and the ill-posed nature of optical imaging reconstructions complicates image interpretation.[38] For these reasons, DOI is often combined with other imaging modalities to minimize distortions and improve reconstructions. DOI has been combined with ultrasound,[32] MRI,[39] PET,[40] and mammography.[36] These approaches provide access to multiple contrast mechanisms that will help in the understanding, validation and clinical acceptance of DOI/DOS. One limitation of these combined approaches is that the spatial distribution of optical contrast is more diffusive (i.e., > radiological boundaries) than that of structure-based contrast.[41] Thus, forcing the optical contrast (i.e., hemoglobin) to fit inside a fixed anatomical region distorts the values of the reconstruction.

While not discussed in detail here, there are other active areas for breast cancer detection involving diffuse signals: (a) fluorescence tomography using exogenous contrast agents, and (b) photoacoustic tomography, with and without exogenous contrast.

Cancer Detection and Characterization

Controversial guidelines recently issued by the U.S. Preventative Services Task Force (USPSTF) concluded that the risk of false positives and complications from biopsies is too high for screening mammography in pre- and peri-menopausal women, i.e. up to age 50.[42] This guideline was based on the poor sensitivity of mammography (approximately 60%) in dense breast tissues. Although the risk of breast cancer is low in this population, women in their forties account for at least a quarter of breast cancer diagnoses and up to 17% of breast cancer deaths each year.[42] The National Cancer Institute estimates that up to 20% of all breast cancers, roughly 40,000 cases per year in the U.S., are not discovered by screening mammography.

Because DOS/DOI measures several functional contrast elements related to cellular metabolism, angiogenesis, and extracellular matrix, optical methods have potential to reduce false positives and negatives in this population. However, the failure of non-model based optical methods (i.e., diaphanography) as a screening tool in the 1980's reminds us of the complexity of improving detection.

Breast lesion detection

Reporting lesion detection rates of DOS/DOI is somewhat misleading because the studies are not performed in a true screening setting. Thus DOS/DOI screening comparison with mammography is premature at best. For this reason, we will not report on sensitivity/specificity values or receiver-operator characteristic (ROC) curves for detection. Keeping this in mind, a recent review of published studies up to 2006 estimated that about 85% of breast lesions are detectable via optical mammography.[43] Whether DOS/DOI can reliably detect mammographically occult lesions is unknown. We will survey the largest population DOS and DOI clinical research since 2005 (i.e., no case studies or instrumentation tests). The studies are challenging to compare as the lesion sizes/types, patient ages/menopausal status, study objectives and measured optical/physiological parameters differed. Despite these difficulties, a consistent story with many subplots has emerged.

A universal finding of these published studies is that ctTHb increases in malignant lesions relative to normal tissues. The ctTHb increase is suspected to be a consequence of tumor angiogenesis. Findings related to ctTHb have been reported in different ways. A time-domain imaging study of 87 carcinomas reported an average ctTHb of 53 ± 32 μM compared with 17.3 ± 6.2 μM in healthy breast tissue.[44] A commercial time-domain instrument reported ctTHb values of 27.7 ± 9.2 and 20.3 ± 4.2 μM for malignant and normal breast ($N = 11$), respectively.[29] Three-dimensional ring reconstructions of malignant breast tumors have reported a mean ctTHb of 48 μM as compared with 13–30 μM in normal tissues ($N = 21$).[45] A more recent three-dimensional parallel plate reconstruction study estimated ~20–30% relative increase in ctTHb between malignant and normal ($N = 41$).[27] A DOS study (without tomography) in 58 carcinomas reported mean ctTHb values of 24.7 ± 9.8 μM and 17.5 ± 7.5 μM in malignant and normal tissues, respectively. Other studies have reported relative ctTHb increases of a few μM using differential absorption ($N = 44$) and response to compression ($N = 36$).[37, 46]

Some groups have reported their findings in images in terms of scoring systems[12,28,33,45] and receiver-operator curves.[27,30,46,47] While some of these studies did not explicitly measure or report ctTHb, the image scores were interpreted in terms of hemoglobin/angiogenic contrast. One study using dynamic compression in 72 cases displayed low sensitivity and specificity; however these lesions were non-palpable (BIRADS 4–5) and no spectroscopic information was applied.[48] The CTLM study in 82 cases showed that the ROC could be improved when optical mammography was used in conjunction with conventional mammography.[30] Important studies about optically-determined ROC curves have been performed.[49]

Detection of lesions by other contrast elements has produced more varied results. It has been long hypothesized that malignant lesions should display lower relative stO_2 as a consequence of increased O_2 consumption/higher metabolism. Although many studies have reported lower stO_2 in malignant lesions, several have not.[4,27,44,50] High resolution spatial imaging of blood vessels may help.[51] Tumor scattering seems to differ from normal although this is challenging to confirm because of scattering-absorption crosstalk. Studies have shown increases in μ_s' [27,44] or alterations in the scattering spectral properties.[4,35,50,52]

Some instruments have indicated additional lesion/normal contrast is available using ctH_2O and ctLipids.[4,10,12,29] In general ctH_2O has been observed to increase and lipids to decrease in malignant lesions. In a 58 tumor DOS study, the changes of malignant versus normal were measured to be $25.9 \pm 13.5\%$ vs. $18.7 \pm 10.3\%$ (ctH_2O) and $58.5 \pm 14.8\%$ vs. $66.1 \pm 10.3\%$ (ctLipid), both in terms of absorption relative to pure subance.[4] Some have not found consistent water differences between malignant/normal, but this may be due to low spectral bandwidth.[47] Water dispositional changes differ between malignant and normal tissues: using a broadband DOS instrument, differences between the bound and free states of water reported by BWI were observed in malignant (1.96 ± 0.3) versus normal (2.77 ± 0.47) tissues ($N = 18$, $p < 0.0001$).[6] The lower BWI suggests that the increases in tumor water are in free rather than bound water.

By using a self-referencing differential spectroscopy method, spectroscopic signatures specific to breast lesions (i.e., "specific tumor components" or STC) have been detected.[14,15] The STC was measured in 40 lesions (malignant, $N = 22$ and benign Fibroadenoma, $N = 18$) and $N = 21$ control subjects. Studies have found that malignant and benign tumors displayed characteristic STC spectral fingerprints, yet normal tissues and control patients had no STC elements. The novelty of this method is that detection via the STC relies on the presence or absence of these spectral fingerprints which report on molecular disposition (i.e., a spectral shape) and not molecular abundance (i.e., a concentration gradient).[15]

Differential diagnosis

The separation of benign from malignant lesions is a challenging and complex problem. Stratification of benign/malignant lesions has been explored by comparing malignant/normal versus benign/normal. The majority of these early studies, which are generally limited in the number of subjects, have been summarized by Leff *et al.*[26]

We note that there are a wide variety of benign lesions, such as defined masses (i.e., Fibroadeoma), liquid sacs (i.e., cysts), and scar tissues (i.e., fibrosis).

Distinctions between benign and malignant lesions have been characterized by increases in ctTHb in malignant/normal that exceed those for benign/normal. This finding is clearest in the ultrasound/DOT instrument that surveyed 73 benign tissues of various types (though only 8 malignant lesions).[32] The majority of benign tissues had significantly lower maximum and average ctTHb: ~38 ± 17.4 µM for benign and 88 ± 24.5 µM for malignant using the average lesion value. Similar results in terms of relative ctTHb ratios and image scores have been found.[12,47] Some studies have found little contrast for fibroadeomas, though limited data exists ($N = 10$).[27] One study found no optical property changes in 4/12 Fibroadenomas, and 7/12 had blood volume increases.[45] Cysts have been detected via alterations in tissue scattering.[12,53,54] There is also some evidence that scattering is different in benign lesions.

In a recently published study, fibroadenomas ($N = 18$) showed STC spectra that are distinct from malignant ($N = 22$) STC spectra.[15] An algorithm successfully characterized the differences between the spectra, and condensed the information into a "malignancy index." The advantage of this approach over previous methods is that it avoids detecting and in this case characterizing tumors by thresholding levels of hemoglobin concentration and/or saturation. While increased levels of hemoglobin are often found in malignant breast tissues, hemoglobin is not a specific marker for tumors. It is expected that other lesion types will present different STC spectra but this has not been confirmed.

Therapeutic Monitoring

Neoadjuvant (i.e., pre-surgical) therapy offers unique opportunities for patient care and cancer drug/technology development.[55] Neoadjuvant therapy often leads to improved breast tissue conservation and avoidance of mastectomy, owing to the reduction of pre-surgical tumor size.[56] The largest potential upside for neoadjuvant chemotherapy is that tumor response to therapy can be assessed *in vivo* on an individual patient basis. Conventional therapeutic end points for cancer treatments are 5–10 year overall survival and disease-free survival. These endpoints offer no feedback during therapy, and lengthen the time until study results are available. Pathological complete response (pCR) is an important therapeutic endpoint that strongly correlates with patient survival.[57] The major advantage of pCR is that assessment is done at the completion of therapy (months) and provides a good projection for survival/ disease free survival (years). However, the assessment of pCR is only done at the conclusion of therapy. Monitoring individual tumor response *during* treatment may provide surrogate end points for therapeutic effectiveness with a goal of not only improving patient overall/disease-free survival, but also improving patient quality of life by minimizing collateral damage from ineffective therapies.

Imaging tumor response to neoadjuvant chemotherapy

If any improvements in patient outcome are to be made, the results of imaging methods must correlate closely with tumor pathologic response. Note that the detection of tumor size changes alone is not conclusive for correlation with pCR; standard clinical tools (physical exam, ultrasound, and mammography) have been shown to be inadequate for predicting tumor pathological response.[58] Functional imaging with advanced techniques such as MRI, Magnetic Resonance Spectroscopy (MRS) and Positron Emission Tomography (PET) has shown improved response assessment capabilities over conventional anatomic imaging techniques.[59-63] However, these functional techniques can be difficult to perform for advanced stage cancer patients (lengthy scan times, exogenous contrast, costly), particularly if frequent measurements are desired. No consensus exists on the optimal time points and the parameters to measure.

DOS and DOI for evaluating tumor response

The first application of NIR optical imaging to track tumor response to neoadjuvant chemotherapy was a 2004 case study using the UC

Irvine DOS instrument.[64] ctTHb and ctH_2O dropped significantly from the pre-treatment baseline by 56% and 67%, respectively over the course of three cycles (three weeks each) of chemotherapy. Importantly, these changes were observed as early as three days post-therapy. The water/lipid ratio and hemoglobin/blood volume changes compared favorably with previous MRI/PET studies.[64] Subsequent case studies supported this initial DOS finding by direct comparison with MRI.[65,66]

The first DOS study that demonstrated physiological endpoints derived from NIR optical signals could be surrogates for pathological response came a few years later.[67] The UC Irvine DOS instrument detected significant differences in optical signals that scaled with final pathological response ($N = 11$). DOS-measured tumor concentrations of ctHHb, ctO_2Hb and ctH_2O dropped $27 \pm 15\%$, $33 \pm 7\%$, and $11 \pm 15\%$, respectively within one week of the first treatment for pathology-confirmed responders ($N = 6$), while non-responders ($N = 5$) and normal side controls showed no significant changes in these parameters. The amount of imaging in this study was limited, but the spectroscopic bandwidth was high and the instrument was a portable/bedside monitor.

DOI studies have supported these initial DOS findings. A combined DOI/Ultrasound technique followed 11 patients, primarily after the second, fourth and final treatment cycles.[68] The authors measured changes in an optical parameter, the "Blood-Volume Index" (BVI) (i.e., lesion volume $\times$ ctTHb). At the completion of treatment, the BVI had dropped significantly: over 60% for pCR or near-pCR, over 50% for partial response, and less than 15% for partial/minor (but visible) responses. In addition, the authors found that the stronger overall responses occurred earlier (cycle 2) and larger than the weaker responders, though the findings were not statistically significant. The study showed a weak correlation between the maximum ctTHb value and the mean vessel density from histological analysis.

A more detailed analysis of DOI imaging procedures by the Dartmouth group compared the optical/physiological responses using different region of interest definitions.[35] The frequency domain tomography instrument (without co-registered MRI) imaged the tumor responses of seven patients, and reported results at the end of two therapy cycles (~four weeks) and just prior to surgery. The authors showed that after four weeks only changes in ctTHb were significantly different between pCR ($N = 4$) and non-pCR ($N = 3$). This result is similar to the previous study,[68] and demonstrated that different methods of ROI selection give different quantitative values but similar trend results.

Using a commercial instrument (SoftScan), Soliman and colleagues recently reported the first detailed multi-time point DOS/DOI on $N = 10$ patients. Measurements were obtained just prior to treatment and at one week, four weeks, eight weeks, and at the conclusion of therapy. A unique feature of this study was that all patients had aggressive disease and received a variety of neoadjuvant treatment regimens. This was done to test the broad applicability of DOS, regardless of tumor cell death mechanism. A single pathologist interpreted all specimens and categorized five subjects as displaying good response and four subjects as minimal or non-responders (none were pCR). When these clinical endpoints were compared to optical endpoints, the responders were clearly separable from non-responders by treatment week four. Changes were dramatic, with drops from baseline of 67.6%, 58.9%, 51.2%, and 52.6% in ctHHb, ctO_2Hb, %ctH_2O, and SP. Corresponding drops in the four non-responders were 17.7%, 18.0%, 15.4%, and 12.6%. Differences between responders and non-responders were statistically significant for all parameters ($p < 0.05$) except for %ctH_2O, which approached significance ($p = 0.0598$). These results are comparable to the UC Irvine 11-patient study cited above.[67]

These initial results, using different instruments and approaches, support for the idea that quantitative NIR functional imaging endpoints can be used to longitudinally monitor and predict clinical treatment response, regardless of the chemotherapeutic strategy, agent, or dose. Patterns of DOS response include an overall decrease in tumor hemoglobin due to drug-induced alterations in tumor cell metabolism and blood vessel density. More specifically, the oxyhemoglobin decrease reflects a diminished vascular supply, while the deoxyhemoglobin drop is representative of a reduction in tumor tissue oxygen consumption that occurs with cell death. Water and scatter power are also sensitive to cell death, and their reduction reflects a progressive loss of cellularity and edema.

What is Next

For the future we see three main areas that need to be addressed: (a) standardization, (b) contrast optimization, and (c) validation. With regard to standardization, it is important from the start that DOS/DOI establish standards for technology, data analysis, and calibration. To date, little has been done but it is likely that commercialization will help address this issue. With regard to contrast optimization, intensive research should continue in developing advanced spectroscopic analysis tools, improving tumor spatial localization, co-registration with conventional radiologic methods, and combining with molecular-targeted exogenous contrast agents. With regard to validation, more expansive multi-center clinical trials are urgently needed. Without these studies, DOS/DOI research will remain promising, but clinically inconclusive.

References

1. Patterson MS, Chance B, *et al.* (1989) Time resolved reflectance and transmittance for the non-invasive measurement of tissue optical properties. *Appl Opt* **28**: 2331–2336.
2. Ishimaru A. (1989) Diffusion of light in turbid material. *Appl Opt* **28**: 2210–2215.
3. Boas DA, O'Leary MA, *et al.* (1997) Detection and characterization of optical inhomogeneities with diffuse photon density waves: A signal-to-noise analysis. *Appl Opt* **36**(1): 75–92.
4. Cerussi A, Shah N, *et al.* (2006) In vivo absorption, scattering, and physiologic properties of 58 malignant breast tumors determined by broadband diffuse optical spectroscopy. *J Biomed Opt* **11**(4): 044005.
5. Ertefai S, Profio AE. (1985) Spectral transmittance and contrast in breast diaphanography. *Med Phys* **12**(4): 393–400.

6. Chung SH, Cerussi AE, *et al.* (2008) In vivo water state measurements in breast cancer using broadband diffuse optical spectroscopy. *Phys Med Biol* **53**(23): 6713–6727.

7. Mourant JR, Fuselier T, *et al.* (1997) Predictions and measurements of scattering and absorption over broad wavelength ranges in tissue phantoms. *Appl Opt* **36**(4): 949–957.

8. Srinivasan S, Pogue BW, *et al.* (2003) Interpreting hemoglobin and water concentration, oxygen saturation, and scattering measured in vivo by near-infrared breast tomography. *Proc Natl Acad Sci USA* **100**(21): 12349–12354.

9. Thomsen S, Tatman D. (1998) Physiological and pathological factors of human breast disease that can influence optical diagnosis. *Ann N Y Acad Sci* **838**: 171-93.

10. Pogue BW, Jiang S, *et al.* (2004) Characterization of hemoglobin, water, and NIR scattering in breast tissue: analysis of intersubject variability and menstrual cycle changes. *J Biomed Opt* **9**(3): 541–552.

11. Cerussi AE, Jakubowski D, *et al.* (2002) Spectroscopy enhances the information content of optical mammography. *J Biomed Opt* **7**(1): 60–71.

12. Taroni P, Torricelli A, *et al.* (2005) Time-resolved optical mammography between 637 and 985 nm: Clinical study on the detection and identification of breast lesions. *Phys Med Biol* **50**(11): 2469–2488.

13. Taroni P, Comelli D, *et al.* (2007) Absorption of collagen: Effects on the estimate of breast composition and related diagnostic implications. *J Biomed Opt* **12**(1): 014021.

14. Kukreti S, Cerussi A, *et al.* (2007) Intrinsic tumor biomarkers revealed by novel double-differential spectroscopic analysis of near-infrared spectra. *J Biomed Opt* **12**(1): 020509.

15. Kukreti S, Cerussi AE, *et al.* (2010) Characterization of metabolic differences between benign and malignant tumors: High-spectral-resolution diffuse optical spectroscopy. *Radiology* **254**(1): 277–284.

16. Wells WA, Wang X, *et al.* (2009) Phase contrast microscopy analysis of breast tissue: Differences in benign vs. malignant epithelium and stroma. *Anal Quant Cytol Histol* **31**(4): 197–207.

17. Garcia-Allende PB, Krishnaswamy V, *et al.* (2009) Automated identification of tumor microscopic morphology based on macroscopically measured scatter signatures. *J Biomed Opt* **14**(3): 034034.

18. Bartrum RJ, Jr., Crow HC. (1984) Transillumination lightscanning to diagnose breast cancer: A feasibility study. *AJR Am J Roentgenol* **142**(2): 409–414.

19. Monsees B, Destouet JM, *et al.* (1988) Light scanning of nonpalpable breast lesions: Reevaluation. *Radiology* **167**(2): 352.

20. Alveryd A, Andersson I, *et al.* (1990) Lightscanning versus mammography for the detection of breast cancer in screening and clinical practice. A Swedish multicenter study. *Cancer* **65**(8): 1671–1677.

21. Profio AE, (1989) Light transport in tissue. *Appl Opt* **28**(12): 2216–2222.

22. Patterson MS, Moulton JD, *et al.* (1991) Frequency-domain reflectance for the determination of the scattering and absorption properties of tissue. *Appl Opt* **30**(31): 4474–4476.

23. Zaccanti G, Bruscaglioni P, *et al.* (1992) Transmission of a pulsed thin light beam through thick turbid media: Experimental results. *Appl Opt* **31**(12): 2141–2147.

24. Fishkin JB, Gratton E. (1993) Propagation of photon-density waves in strongly scattering media containing an absorbing semi-infinite plane bounded by a straight edge. *J Opt Soc Am A* **10**(1): 127–140.

25. Tromberg BJ, Svaasand LO, *et al.* (1993) Properties of photon density waves in multiple-scattering media. *Appl Opt* **32**(4): 607–616.

26. Leff DR, Warren OJ, *et al.* (2008) Diffuse optical imaging of the healthy and diseased breast: A systematic review. *Breast Cancer Res Treat* **108**(1): pp. 9–22.

27. Choe R, Konecky SD, *et al.* (2009) Differentiation of benign and malignant breast tumors by in-vivo three-dimensional parallel-plate diffuse optical tomography. *J Biomed Opt* **14**(2): 024020.

28. Fournier LS, Vanel D, *et al.* (2009) Dynamic optical breast imaging: A novel technique to detect and characterize tumor vessels, *Eur J Radiol* **69**(1): 43–49.

29. Intes, X. (2005) Time-domain optical mammography SoftScan: Initial results. *Acad Radiol* **12**(8): 934–947.

30. Poellinger A, Martin JC, *et al.* (2008) Near-infrared laser computed tomography of the breast first clinical experience. *Acad Radiol* **15**(12): 1545–1553.

31. van de Ven SM, Elias SG, *et al.* (2009) Diffuse optical tomography of the breast: preliminary findings of a new prototype and comparison with magnetic resonance imaging. *Eur Radiol* **19**(5): 1108–1113.

32. Zhu Q, Cronin EB, *et al.* (2005) Benign versus malignant breast masses: Optical differentiation with US-guided optical imaging reconstruction. *Radiology* **237**(1): 57–66.

33. Grosenick D, Moesta KT, *et al.* (2005) Time-domain scanning optical mammography: I. Recording and assessment of mammograms of 154 patients. *Phys Med Biol* **50**(11): 2429–2449.

34. Yates T, Hebden JC, *et al.* (2005) Optical tomography of the breast using a multi-channel time-resolved imager. *Phys Med Biol* **50**(11): 2503–2517.

35. Jiang S, Pogue BW, *et al.* (2009) Evaluation of breast tumor response to neoadjuvant chemotherapy with tomographic diffuse optical spectroscopy: Case studies of tumor region-of-interest changes. *Radiology* **252**(2): 551–560.

36. Fang Q, Carp SA, *et al.* (2009) Combined optical imaging and mammography of the healthy breast: Optical contrast derived from breast structure and compression. *IEEE Trans Med Imaging* **28**(1): 30–42.

37. Xu RX, Young DC, *et al.* (2007) A prospective pilot clinical trial evaluating the utility of a dynamic near-infrared imaging device for characterizing suspicious breast lesions. *Breast Cancer Res* **9**(6): pp. R88.

38. Li A, Boverman G, *et al.* (2005) Optimal linear inverse solution with multiple priors in diffuse optical tomography. *Appl Opt* **44**(10): 1948–1956.

39. Brooksby B, Pogue BW, *et al.* (2006) Imaging breast adipose and fibroglandular tissue molecular signatures by using hybrid MRI-guided near-infrared spectral tomography. *Proc Natl Acad Sci USA* **103**(23): 8828–8833.

40. Konecky SD, Choe R, *et al.* (2008) Comparison of diffuse optical tomography of human breast with whole-body and breast-only positron emission tomography. *Med Phys* **35**(2): 446–455.

41. Li A, Liu J, *et al.* (2008) Assessing the spatial extent of breast tumor intrinsic optical contrast using ultrasound and diffuse optical spectroscopy. *J Biomed Opt* **13**(3): pp. 030504.

42. USPSTF (2009) Screening for breast cancer: U.S. Preventive Services Task Force recommendation statement. *Ann Intern Med* **151**(10): 716–726, W-236.

43. Enfield LC, Gibson AP, *et al.* (2009) Optical tomography of breast cancer-monitoring response to primary medical therapy. *Target Oncol* **4**(3): 219–233.

44. Grosenick D, Wabnitz H, *et al.* (2005) Time-domain scanning optical mammography: II. Optical properties and tissue parameters of 87 carcinomas. *Phys Med Biol* **50**(11): 2451–2468.

45. Enfield LC, Gibson AP, *et al.* (2007) Three-dimensional time-resolved optical mammography of the uncompressed breast. *Appl Opt* **46**(17): 3628–3638.

46. Chance B, Nioka S, *et al.* (2005) Breast cancer detection based on incremental biochemical and physiological properties of breast cancers: a six-year, two-site study. *Acad Radiol* **12**(8): 925–933.

47. Poplack SP, Tosteson TD, *et al.* (2007) Electromagnetic breast imaging: Results of a pilot study in women with abnormal mammograms. *Radiology* **243**(2): 350–359.

48. Athanasiou A, Vanel D, *et al.* (2007) Optical mammography: A new technique for visualizing breast lesions in women presenting non palpable BIRADS 4–5 imaging findings: Preliminary results with radiologic-pathologic correlation. *Cancer Imaging* **7**: 34–40.

49. Pogue BW, Davis SC, *et al.* (2006) Image analysis methods for diffuse optical tomography. *J Biomed Opt* **11**(3): pp. 33001.

50. Spinelli L, Torricelli A, *et al.* (2005) Characterization of female breast lesions from multi-wavelength time-resolved optical mammography. *Phys Med Biol* **50**(11): 2489–2502.

51. Heffer E, Pera V, *et al.* (2004) Near-infrared imaging of the human breast: Complementing hemoglobin concentration maps with oxygenation images. *J Biomed Opt* **9**(6): 1152–1160.

52. Li C, Grobmyer SR, *et al.* (2008) Noninvasive in vivo tomographic optical imaging of cellular morphology in the breast: Possible convergence of microscopic pathology and macroscopic radiology. *Med Phys* **35**(6): 2493–2501.

53. van de Ven S, Elias S, *et al.* (2009) Diffuse optical tomography of the breast: initial validation in benign cysts. *Mol Imaging Biol* **11**(2): 64–70.

54. Gu X, Zhang Q, *et al.* (2004) Differentiation of cysts from solid tumors in the breast with diffuse optical tomography. *Acad Radiol* **11**(1): 53–60.

55. Wolff AC, Berry D, *et al.* (2008) Research issues affecting preoperative systemic therapy for operable breast cancer. *J Clin Oncol* **26**(5): 806–813.

56. Fisher B, Brown A, *et al.* (1997) Effect of preoperative chemotherapy on local-regional disease in women with operable breast cancer: Findings from National Surgical Adjuvant Breast and Bowel Project B-18. *J Clin Oncol* **15**(7): 2483–2493.

57. Fisher B, Bryant J, *et al.* (1998) Effect of preoperative chemotherapy on the outcome of women with operable breast cancer. *J Clin Oncol* **16**(8): 2672–2685.

58. Yeh E, Slanetz P, *et al.* (2005) Prospective comparison of mammography, sonography, and MRI in patients undergoing neoadjuvant chemotherapy for palpable breast cancer. *AJR Am J Roentgenol* **184**(3): 868–877.

59. Meisamy S, Bolan PJ, *et al.* (2004) Neoadjuvant chemotherapy of locally advanced breast cancer: Predicting response with in vivo (1)H MR spectroscopy — a pilot study at 4 T. *Radiology* **233**(2): 424–431.

60. Duch J, Fuster D, *et al.* (2009) (18)F-FDG PET/CT for early prediction of response to neoadjuvant chemotherapy in breast cancer. *Eur J Nucl Med Mol Imaging* **36**(10): 1551–1557.

61. Baek HM, Chen JH, *et al.* (2009) Predicting pathologic response to neoadjuvant chemotherapy in breast cancer by using MR imaging and quantitative 1H MR spectroscopy. *Radiology* **251**(3): 653–662.

62. Padhani AR, Hayes C, *et al.* (2006) Prediction of clinicopathologic response of breast cancer to primary chemotherapy at contrast-enhanced MR imaging: Initial clinical results. *Radiology* **239**(2): 361–374.

63. Mankoff DA, Dunnwald LK, *et al.* (2002) Blood flow and metabolism in locally advanced breast cancer: Relationship to response to therapy. *J Nucl Med* **43**(4): 500–509.

64. Jakubowski DB, Cerussi AE, *et al.* (2004) Monitoring neoadjuvant chemotherapy in breast cancer using quantitative diffuse optical spectroscopy: A case study, *J Biomed Opt* **9**(1): 230–238.

65. Shah N, Gibbs J, *et al.* (2005) Combined diffuse optical spectroscopy and contrast-enhanced magnetic resonance imaging for monitoring breast cancer neoadjuvant chemotherapy: A case study. *J Biomed Opt* **10**(5): pp. 51503.

66. Choe R, Corlu A, *et al.* (2005) Diffuse optical tomography of breast cancer during neoadjuvant chemotherapy: A case study with comparison to MRI. *Med Phys* **32**(4): 1128–1139.

67. Cerussi A, Hsiang D, *et al.* (2007) Predicting response to breast cancer neoadjuvant chemotherapy using diffuse optical spectroscopy. *Proc Natl Acad Sci USA* **104**(10): 4014–4019.

68. Zhu Q, Tannenbaum S, *et al.* (2008) Noninvasive monitoring of breast cancer during neoadjuvant chemotherapy using optical tomography with ultrasound localization. *Neoplasia* **10**(10): 1028–1040.

69. Franceschini MA, Moesta KT, *et al.* (1997) Frequency-domain techniques enhance optical mammography: Initial clinical results. *Proc Natl Acad Sci USA* **94**(12): 6468–6473.

Chapter 8

OCT for Skin Cancer

Gordon McKenzie[*,†] *and Adam Meekings*[*,‡]

Introduction

OCT has transformed ophthalmology over the last ten years, going from a niche product used by only a few specialists to a gold standard imaging tool with tens of papers per month and more than 10,000 installed units in hospitals and optometry clinics worldwide. Like ultrasound, it provides cross-sectional images of tissue, but as it is based on optics rather than sound, it provides far greater resolution than possible with conventional ultrasound systems.

The greatest challenges to the introduction of OCT into clinical practice are

- Image quality
- Clinical understanding of the images
- Clinical implications of its use
- Commercial model that would support uptake

While the commercial models that might support the uptake of OCT are probably outside the scope of a chapter such as this, some simple technical aspects of the image quality, and the clinical questions currently being addressed are both highly relevant and will be examined here.

---

[*] Michelson Diagnostics Ltd, Orpington, UK
[†] Medical Applications Director
[‡] Clinical Imaging Scientist

OCT for dermatology is a rapidly expanding field, academically active and understanding is growing exponentially. For that reason, this chapter can hope only to provide a useful background for the reader, being as it is a snapshot of the field in 2010.

What is OCT?

OCT is a form of "laser ultrasound" that involves shining a low powered infrared beam into the tissue and collecting photons that are reflected back by the structures beneath the surface. OCT was independently conceived by Tanno and Chiba of Yamagata University in 1990,[1] and by Huang and Fujimoto at MIT in 1991,[2] the two groups working entirely independently from each other, but the basic concept goes right back to 1887 when Michelson and Morley invented the interferometer.

OCT has some key advantages over other imaging technologies: the absence of any ionizing radiation or carcinogenic side effects, the easy applicability, the relatively low investment costs (compared to computed tomography and magnetic resonance tomography) and the relatively short examination time of only a few minutes.

With the help of OCT, information may be obtained in a non-invasive manner on the extension, the structure, and the possible dignity (benign or malignant) of such lesions without the need to take an invasive biopsy (3D mapping). For example, if the distinction between dysplasia and early invasive carcinoma were possible, the subsequent treatment, for example the kind of resection, PDT or other forms of local therapy, could be better planned. OCT would offer a non-invasive examination tool, and would provide quasi-histological information about the tissue, the so-called optical biopsy, and would give new innovative possibilities in the field of diagnostics.

Basic concepts in OCT

OCT builds on the same basic model as a conventional ultrasound system, with infrared light replacing an acoustic wave. Light is shone into the tissue, and the time taken for photons to reflect from structures within the skin and come back to a detector is measured and translated into a distance travelled. Hence, by tightly focusing the beam into a narrow column, and by scanning the beam laterally through the tissue, a map can be built up of the structures below the surface.

Consequences

The substitution of light for sound has two important consequences. Compared to ultrasound, much higher resolution is achieved: in the case of VivoSight, better than 10 microns in all axes compared to perhaps 200 microns for a typical high-frequency ultrasound system. Unfortunately, this additional resolution comes at a price of reduced imaging depth, typically between 1 and 1.5 mm in skin at the 1,300 nm wavelength typically used. This is not a technical limitation, but rather a function of the fundamental principle of OCT — the detection of photons that have reflected from a structure and then travelled back out of the tissue without any further interactions. These 'ballistic' photons become more and more rare, and beyond a certain depth there are simply no ballistic photons to be detected.

This attenuation of signal has been particularly challenging in the imaging of skin. The earliest successful systems concentrated on the eye because of a perfect synergy between technical capability and clinical interest. The eye is a fundamentally nice place to image. The optical path is non-scattering, and the retina is optically flat, thin, and partially transparent. Skin, by contrast, is dense, highly reflecting at the surface and tends not to be flat — altogether a more challenging environment in which to image.

The development of practical OCT systems

Early OCT systems operated in the time domain. Here, the mechanical movement of the reference arm mirror selected the depth at which an image was being acquired. As light sources improved, giving a broader spectrum of light and therefore higher axial resolution, time domain OCT allowed for extremely high axial resolution images to be collected. By coupling the point of optical focus to the reference mirror mechanism, a dynamic focus could be established that tracked the point of acquisition, achieving high lateral focus. In this manner, OCT became a powerful imaging tool for the laboratory, and the number of publications grew steadily; eventually the OCT sessions became the largest at the SPIE Photonics West conference in California each year.

Technical limitations

The principal limitation of this first generation of technology was imaging speed. Time domain systems suffer from a fundamental signal to noise problem: they collect data one pixel at a time, meaning that at any particular instant, more than 99% of the light put into the system is not being used to form an image. At slow speeds this is not too much of a problem as there is plenty of light, but as images are collected more quickly, then the time available for each pixel is reduced until signal to noise becomes a major problem. In early systems this resulted in image capture times of multiple seconds per frame — too slow for practical use *in vivo* in dermatology.

This speed limitation was addressed by Fercher's group in the mid '90s.[3,4] They realized that the interferogram could be collected in the frequency domain (FD-OCT), allowing the entire scan line to be collected simultaneously, rather than requiring a mirror mechanism to scan along the line. This means that the majority of the light that reflects balistically can be used as signal rather than being discarded. Building on this development, advances in the light sources and detectors have since then enabled very rapid image capture.

FD-OCT has one significant disadvantage: it is no longer possible to achieve high lateral resolutions by dynamic focusing: there is no longer a moving mirror to track. This means that for a given desired depth of field, there is a fixed limit on the lateral resolution.

OCT is an unusual imaging technique in that the axial resolution and lateral resolution are obtained by entirely different methods, and can vary independently of each other within the same

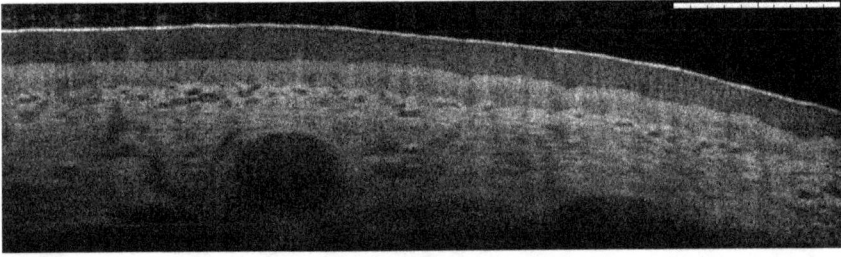

Fig. 1. *In vivo* OCT image of skin, showing the keratinized layer, stratum corneum, epidermis, dermis, capillary networks and the top of large blood vessels. Scale bar is 1 mm.

system. Whilst axial resolution is determined by light source properties, the lateral resolution is a function of the desired depth of focus. A 1 mm depth of focus implies a minimum Gaussian beam radius of 17.5 µm (equivalent to 20.6 µm FWHM). For dermatology, this is often too coarse for useful clinical images.

The 2010 state of the art in clinical OCT for dermatology

To overcome this fundamental limitation of single-beam FD-OCT systems, we have developed a solution in which we simultaneously scan multiple beams, focused at slightly different depths, and compile an image mosaic from the resulting multiple FD-OCT interferograms. Each sub-beam has a theoretical FWHM diameter of just 10.3 µm,[5] and practical measurements by the National Physical Laboratory have shown the implemented system to be capable of resolving 7.5 µm laterally.[6] The resulting images are excellent, with an added bonus that the smaller spot size improves contrast. A typical image of skin is shown in Fig. 1.

This multi-beam system has been implemented in the Michelson Diagnostics VivoSight™ OCT system.

OCT in Comparison to Other Imaging Techniques

Imaging does not currently have much influence on clinical dermatology. Visual inspection is still the mainstay of clinical practice, and with extensive training and experience, senior dermatologists show excellent performance in the diagnosis of skin cancers.

Dermatoscope

A dermatoscope is a low powered magnifying glass with specific illumination is used to allow a higher resolution view of the surface of the skin. Used by a skilled clinician, the dermatoscope can improve diagnosis of pigmented lesions, but a high level of training is required before this improvement can be realized.

Ultrasound

Ultrasound has not found widespread use in clinical practice. Very high frequency ultrasound systems can offer quite good resolution, but a standard HFUS system rarely offers better than 150–200 micron resolution laterally.[7]

On ultrasound systems, skin tumors usually manifest as homogeneously echo-free areas in contrast to their surrounding tissue.[8]

In a recently published study, Marmur *et al.* found no significant difference between the ability of HFUS and clinical visualization in judging the size of NMSC lesions prior to Mohs surgery,[7] although another study did find HFUS to be capable of predicting local recurrence after PDT at 1 year.[9] Another study concluded that, for tumor thickness measurement, OCT was "more accurate and less biased than HFUS", finding that both modalities were prone to some overestimate of thickness, but HFUS more so than OCT.[10]

These factors, coupled with practical difficulties in taking measurements, and in translating the margins observed on the screen to a point on the skin, take up of ultrasound in dermatology has so far been very low.

Confocal imaging

Confocal microscopy has been shown in numerous studies to have excellent diagnostic capabilities when applied to skin cancer. Its cellular resolution gives very good specificity, but the narrow field of view, limited depth of imaging (approximately 200 microns) and relatively slow image acquisition speed have limited the clinical uptake of systems.

Another challenge has been the interpretation of the images. The very high resolution means that the images have more in common with histology than other imaging techniques, but the en-face orientation and the differing chromophores mean that significant training is necessary to be able to make use of the information presented.

To combat this challenge, Lucid Inc., a leading producer of *in vivo* confocal microscope systems, have recently introduced the VivaNet system. Here, the images are acquired locally then uploaded to a central system where a highly trained observer will analyze the image and present a diagnosis back to the clinician. This is an interesting business model on several levels. First, it addresses the problem of training of end users, meaning that a nurse or technician can scan the patient and then present the results to the dermatologist having had an expert provide a definitive diagnosis. Second, it provides a means for the business to provide the system on a pay per use basis, reducing the economic and training hurdles to adoption of the technology. The clinic pays as they use the system, and in effect only pay when they realize a benefit from the technology.

Morphology of the Skin When Examined by OCT

Normal skin with no lesions scanned using OCT presents as a layered architecture with variations in the tone indicating individual layers and complex structures.

Stratum corneum

The initial layer of the skin, the Stratum Corneum, presents as a thick hyporeflective layer below the hyperreflective boundary skin surface.

This hyperreflective layer has often been misinterpreted as the Stratum Corneum itself; however, the Stratum Corneum contains the majority of the coiled sudoriferous duct (sweat gland duct) structure compared to the lower Stratum Spinosum layer (as described in *Gray's Anatomy of the Human Body* Fig. 940). Therefore the Stratum Corneum layer is, as shown in Fig. 2, the layer above the broken line.

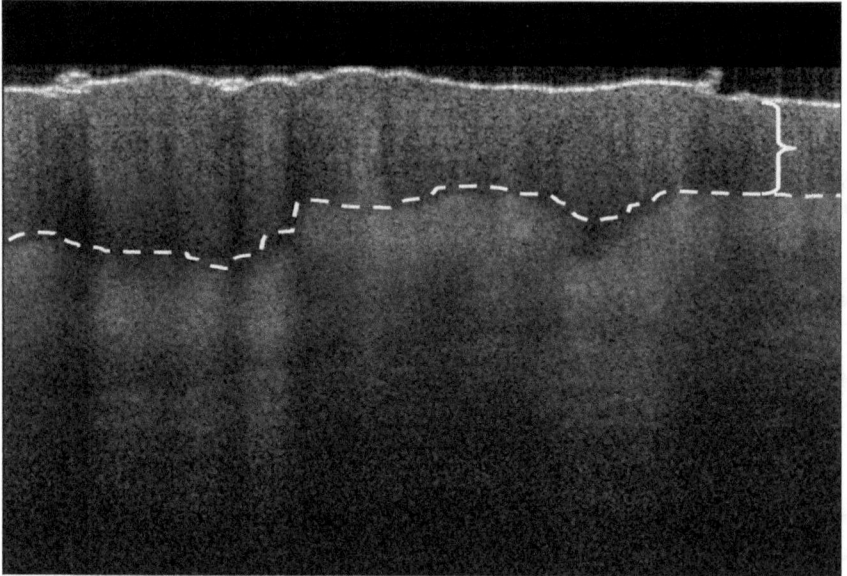

Fig. 2. OCT scan of epidermis and dermis of glabrous skin with the Stratum Corneum boundary indicated.

The Stratum Corneum is identifiable in skin of tactile regions, the sole of the foot and palm. The layer is recognizable in glabrous skin; however, it is unable to be resolved in follicular regions due to its substantially decreased thickness (Fig. 3).

In regions of thick Stratum Corneum the dermal-epidermal junction interface is less defined.

Stratum Corneum thickness varies in depth from an average of 0.47 mm on the fingertip to less than 0.02 mm on the wrists, forearms and other follicular skin regions. The Stratum Corneum as previously stated can be defined by the presence of sudoriferous ducts/sweat ducts traversing the layer (Fig. 4).

The sudoriferous gland ducts, sweat ducts, present as hyper-reflective coiling structures spanning the Stratum Corneum commonly at an incline to the surface layer of the epidermis. The ducts can only be resolved in skin regions with substantially thick Stratum Corneum layers. The reflective properties of the ducts also

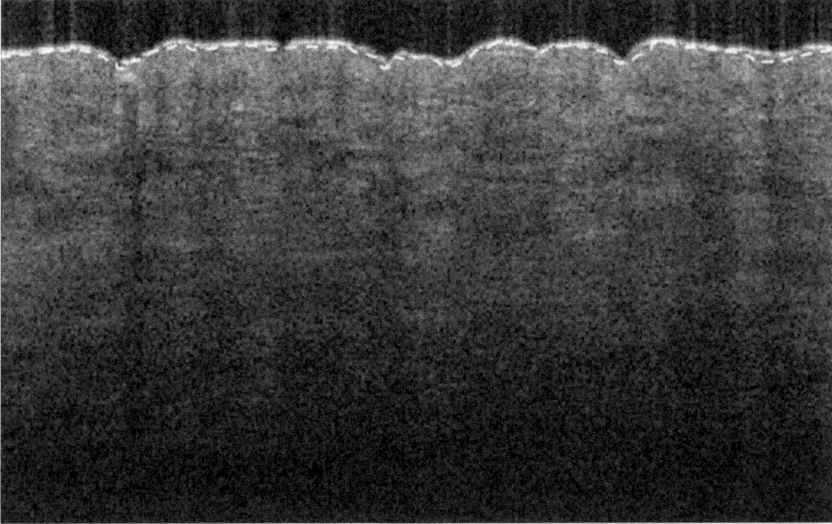

Fig. 3. OCT scan of epidermis and dermis of the wrist showing a reduction in Stratum Corneum depth to the point at which it is unresolvable. Stratum Spinosum boundary is highlighted.

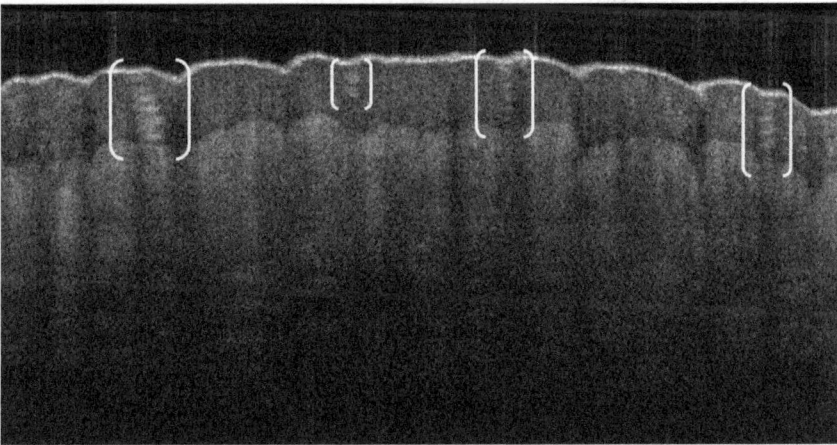

Fig. 4. OCT scan of the epidermis and dermis of the fingertip highlighting the sudoriferous gland duct.

cause a shadow effect to appear obscuring features in the lower layers of the skin.

Stratum Spinosum

The Stratum Spinosum is the next fully resolvable epidermal layer after the Stratum Corneum presenting as hyperreflective in comparison to the hyporeflective S. Corneum. The Stratum Spinosum is identifiable in OCT scans of all anatomical regions; however, a definitive boundary between the Stratum Spinosum and the papillary dermis layer cannot always be differentiated (Figs. 5 and 6).

As with the Stratum Corneum layer the Stratum Spinosum layer is thinner in follicular skin compared to glabrous skin.

In some skin regions, primarily those with thickened S. Corneum, the Stratum Granulosum is identifiable. The S. Granulosum presents as a hyporeflective layer at the boundary of the Stratum Corneum and the S. Spinosum. In some instances the S. Granulosum band may not be visible along the full Stratum Corneum–Stratum Spinosum boundary but show as intermittent hyporeflective darkened regions distinguishable from the speckle on the scan.

The interface between the Stratum Spinosum/Basale layers and the papillary dermis can be identified by the change in scatter pattern from the uniform light grey of the Stratum Spinosum to the less granular appearing papillary dermis, which is more evident within skin regions possessing a thickened S. Corneum; furthermore the presence of rete pegs and ridges also provide a distinct boundary indicating the dermal epidermal junction.

At the site of interaction between the Stratum Basale (basal membrane layer of the epidermis) and the papillary dermis the ability to identify differentiated layers and structures is reduced. Within this region it is possible to identify blood vessel networks of varying size.

Papillary dermis

The papillary dermis including the rete ridges is the initial layer of the dermis. While less identifiable in glabrous skin; the papillary

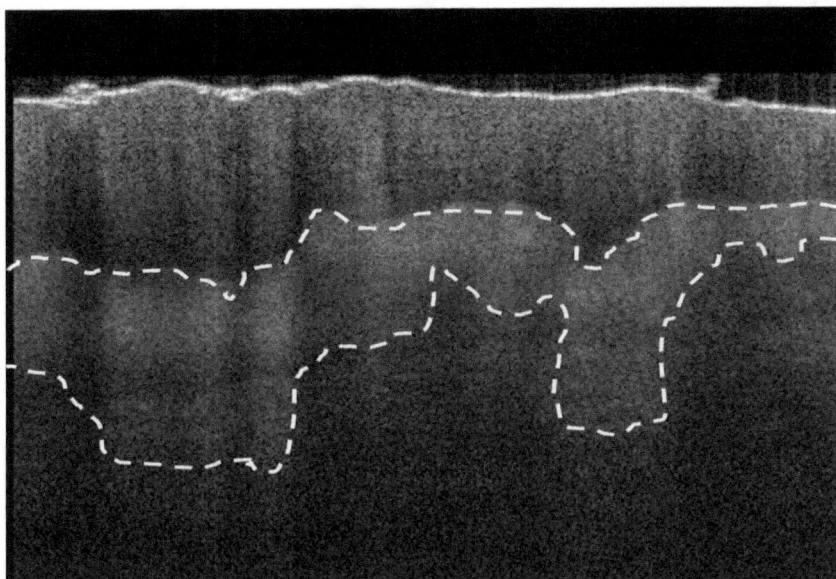

Fig. 5. OCT scan of the epidermis and dermis of glabrous skin with the Stratum Spinosum layer indicated.

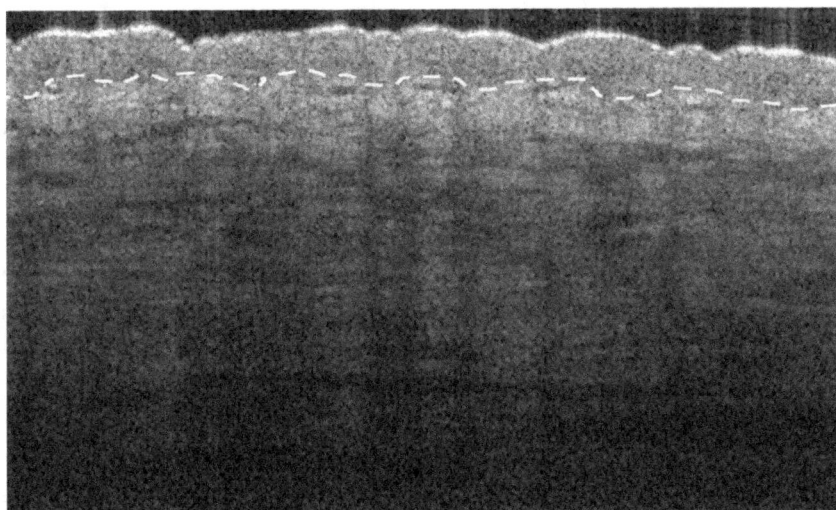

Fig. 6. OCT scan of epidermis and dermis of follicular skin with the Stratum Spinosum layer indicated.

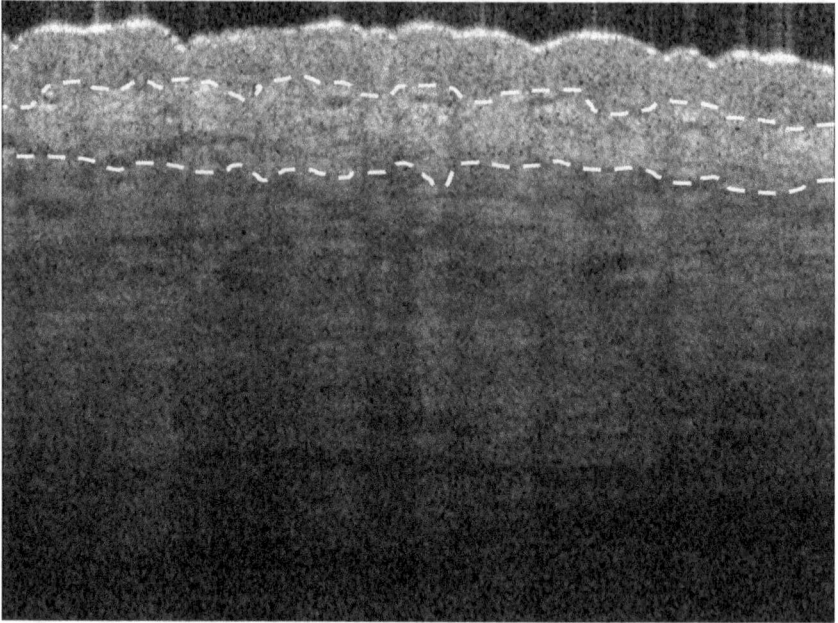

Fig. 7. OCT scan of follicular skin indicating the papillary dermis layer and its boundary with the Stratum Spinosum.

dermis appears as a more hyperreflective layer in comparison to the Stratum Spinosum in follicular skin. Conversely at the interface of the papillary dermis with the S. Spinosum, the region containing the rete pegs, the papillary dermis has a darker scatter pattern (Fig. 7). This area of rete pegs can readily be identified in palmoplantar regions where they occur in relation to surface contours such as the finger print ridges, but can be less identifiable in follicular regions of the skin.

Reticular dermis

The reticular dermis in the bulk of anatomical regions comprises the majority of the dermis depth. In contrast to the hyperreflective papillary dermis the reticular dermis presents as a predominantly hyporeflective region (Fig. 8). The difference in scatter patterns

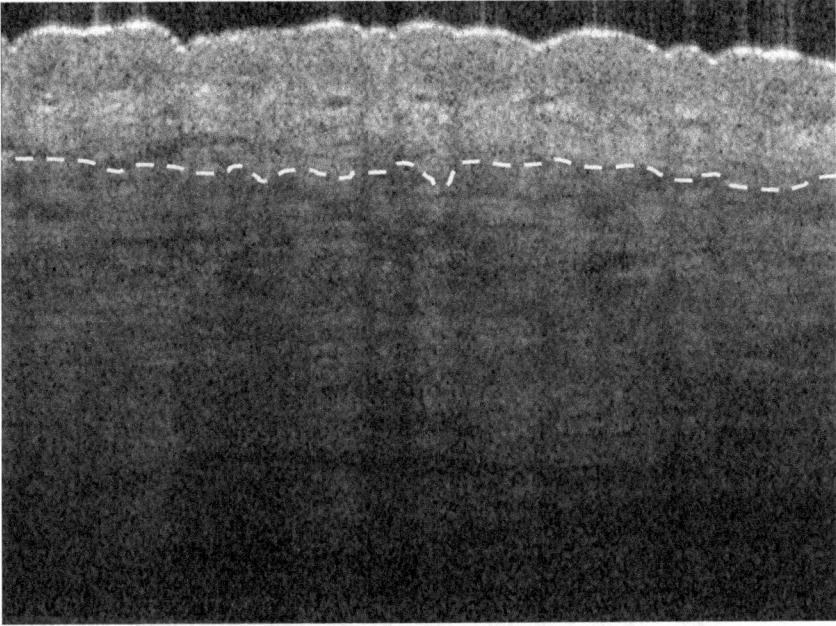

Fig. 8. OCT scan of follicular skin indicating the boundary between the papillary and reticular layers of the dermis.

between the dermis layers is presumably due to the variation in arrangement and density of collagen fibres in the respective regions.

Identifiable structures within the epidermis and dermis

Hair follicles and to some extent the hair bulbs can be identified within the epidermis and dermis respectively. The hair bulb located mainly within the reticular dermis presents as a hyporeflective bulbous structure consisting of the hair bulb and adjacent sebaceous glands. The hair follicle protruding from the bulb often presents as a hyporeflective structure traversing the epidermal layers and penetrating the S. Corneum (Fig. 9). The hair follicle after leaving the Stratum Corneum appears to become more hyperreflective and in doing so, similar to the sweat ducts, causes a shadow to be cast on the skin layers directly beneath the area of highest reflectivity.

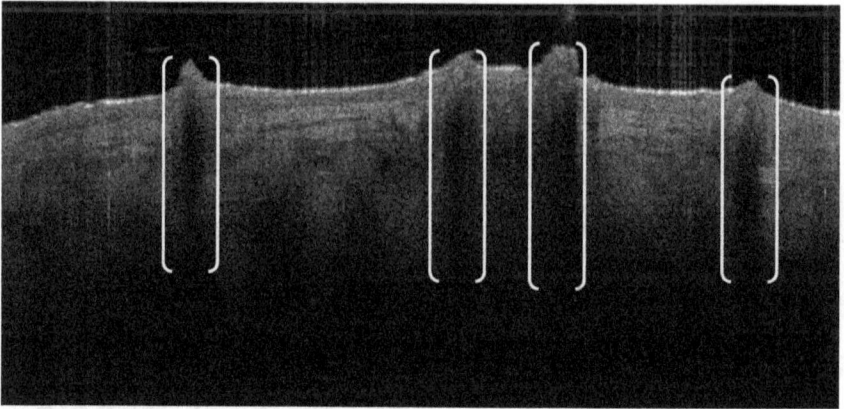

Fig. 9. OCT scan of epidermis and dermis of follicular skin showing the hair bulb and hair follicle.

Blood vessels in some skin regions scanned can be identifiable as hyporeflective tube like structures often appearing with tapering ends located within the dermis layers, primarily the reticular dermis layer (Fig. 10). Size of the blood vessels differs between areas scanned, becoming more prominent in areas of the skin overlying striated muscle and skin with thin S. Corneum.

OCT for Skin Cancer

Early papers on OCT tended to be highly technical. It was challenging to scan tissue *in vivo*, so studies often worked on phantoms, models or on *in vitro* samples. Second, those few papers that did try OCT *in vivo* for skin tended to conclude that the results showed promise but further work was needed to make it a practical clinical tool. For these reasons, this review concentrates on recent papers where the technology is more capable of realizing clinically useful images. It should be noted that the OCT field is incredible active, and it is certain that this review misses more than it covers. For those with an active interest in this field, Eric Swanson's 'OCT News' service is highly recommended.

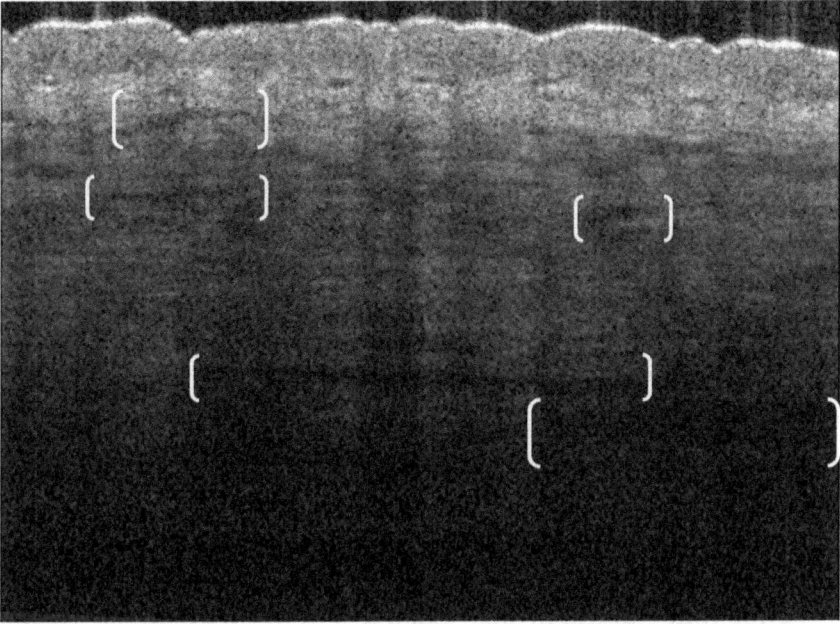

Fig. 10. OCT scan of epidermis and dermis of follicular skin with the blood vessels indicated.

Relevance to non-melanoma skin cancer

Basal Cell Carcinoma (BCC) is the most common non-melanoma skin cancer.[11] The epithelial component consists of solid masses of uniform cells resembling the epidermal structure joined to the lower surface of the epidermis in multiple places. In histological sections, fixation retraction spaces are regularly present in between the tumor masses and the surrounding fibrous stroma, which contains inflammatory cells and dilated vessels.[12]

OCT is capable of visualizing altered skin architecture and histopathological correlates of BCC, allowing accurate differentiation of BCC tumor from normal tissue.[13] It has not yet been proven that OCT could be used clinically for the differentiation of BCC subtypes,[14] but differentiation has been proven in other areas,[15] and the monitoring of inflammation appears to work well,[16] so this may be possible in time. Olmedo showed in a pilot study that "sites matched the histologic features seen on light microscopy, with excellent correlation between optical coherence tomographic images and histopathologic features of superficial, nodular, micronodular and infiltrative basal cell carcinomas".[17] He also showed in a separate study that the thickness of BCC could be reliably measured using OCT to a depth of approximately 1 mm,[18] correlating well with the imaging depth observed by the author in his own work.

Morphology of basal cell carcinoma in OCT images

The nodular basal cell carcinoma presents as a darkened cell mass/basiloid nest in comparison to the surrounding layers. The basiloid nest in this case is primarily ovoid in shape with no distinguishable lobules budding from the central mass (Fig. 11).

Boundaries between the carcinoma and surrounding cell layers are evident due to the presence of a hyporeflective halo-like region between the basiloid nest and the stroma of the surrounding tissue. The hyporeflective area of the nest periphery could be due to the presence of peripheral palisading occuring and the high nuclear material content causing hyporeflectivity.

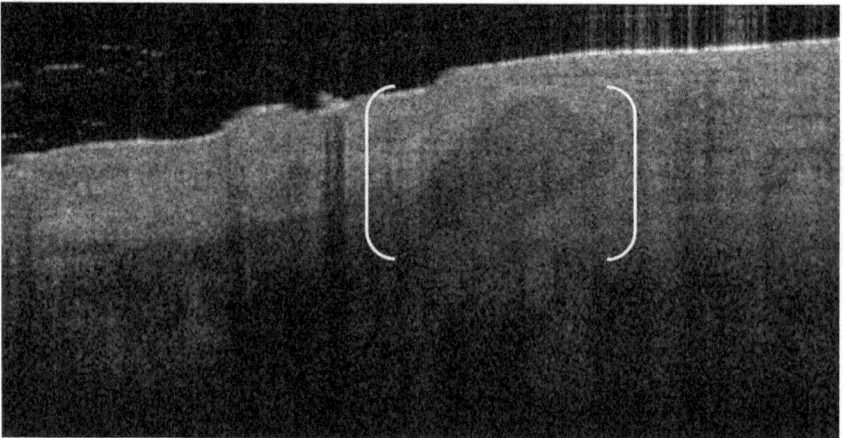

Fig. 11. OCT scan of a nodular basal cell carcinoma.

The presence of the nest has caused disruption to the normal undulating layer of the Stratum Spinosum causing greater protrusion into the dermis region. Rete ridges and other interfaces between the Stratum Spinosum and papillary dermis are not distinguishable in the areas surrounding the basiloid nest.

The superficial basal cell carcinoma possesses a well-defined boundary, as a result of the peripheral palisading region surrounding the basiloid nest. The stroma surrounding the nest, primarily in the epidermis facing region, more hyperreflective in comparison to the adjacent tissue (Fig. 12).

The hyporeflective palisading region between the basiloid nest and the stoma is larger in the lower regions of the carcinoma compared to the regions closer to the epidermis. Perfusion of the basiloid nest by one or more blood vessels directed to the carcinoma site is evident. Disruption to the traditional undulating pattern of the Stratum Spinosum papillary dermis boundary results in the interface not being identifiable in regions directly under the basiloid nest; neither are pronounced rete pegs present.

The superficial multifocal basal cell carcinoma possesses no identifiable hyporeflective palisading region between the basiloid cell nest and the stroma of the surrounding tissue; however, a

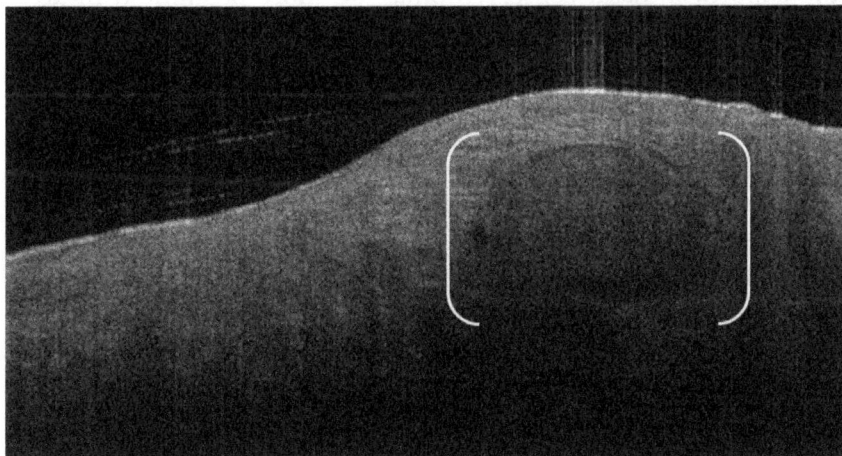

Fig. 12. OCT scan of a superficial basal cell carcinoma.

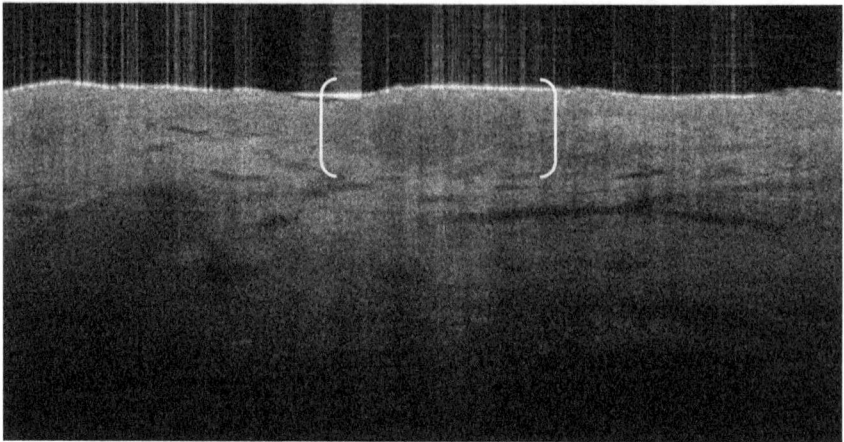

Fig. 13.   OCT scan of a superficial multifocal basal cell carcinoma.

significant difference in the scattering pattern of the nest compared to the surrounding cells is still noticeable (Fig. 13).

The basal cell carcinoma appears to consist of multiple basiloid nests of varying sizes, but follow the same pattern of ovoid shaped structures.

A large blood vessel can be seen linking to the basiloid nest group, meaning that connecting blood vessel networks to nodular and superficial basal cell carcinomas are more prominent in areas where more progressed carcinomas exist.

Discussion and conclusions

Nodular and superficial basal cell carcinomas (BCC) are identifiable through scans by optical coherence tomography. Scans of suspected nodular and superficial BCC identify nests of cells that do not share the common scatter pattern as their surrounding tissue layers, normally presenting as hyporeflective groups of cells.

Nodular and superficial basal cell carcinomas are distinguished mainly by the area the basiloid nest is present. Nodular BCC have nests that are present within the upper dermis layers whereas superficial BCC nest are contained within the lower part of the epidermis and superficial dermis layers.

The appearance of the nodular and superficial basal cell carcinoma appears to remain consistent through multiple samples scanned, appearing as variations of spherical or ovoid structures.

OCT in some instances of superficial and nodular BCC can identify when present the border between the basiloid cell nest and the stroma of the surrounding tissue. This is identifiable not only through the difference in scatter pattern of the carcinoma tissue, but also due to the presence of the palisading region between the basiloid nest and the stroma; recognised as a hyporeflective periphery to the basiloid nest. The hyporeflective periphery is usually more noticeable in the edge of the basiloid nest facing the dermis compared to the side facing the epidermis. The hyporeflective boundary is a result of peripheral palisading of the basaloid nest and increased nuclear content.

Mucin is a glycosylated protein that exists as a gel secretion from tissues; in the case of most carcinomas, overexpression of mucin is associated with the presence of a carcinoma.

Blood vessel directed growth and perfusion of the carcinoma nests is not constant in all nodular or superficial basal cell carcinomas and appears to be more prominent where large groups of basiloid nests are present. Areas of nests that have a large network of identifiable blood vessels often means that the carcinoma in that area is either progressed to an advanced stage or presents as multiple joining basiloid nest islands over an area.

In all cases identified as nodular or superficial basal cell carcinomas thus far the basiloid nest disrupts the regular undulating appearance of the Stratum Spinosum layer of the epidermis and in most cases the rete pegs of the interface between the Stratum Spinosum and the papillary dermis are not identifiable.

While it still seems a little early to try and apply OCT for diagnostic purposes, early clinical work has identified a number of features that appear to be common to the majority of nodular and super-ficial basal cell carcinomas:

- Noticeable change in scatter pattern presenting as more hyporeflective to adjacent tissue layers

- Spherical or ovoid shaped nest
- Disruption to the undulating pattern of the Stratum Spinosum layer including loss of form resulting in the interface between it and the papillary dermis becoming unrecognisable

The following features are not able to identify nodular or superficial BCC, but rather are able to confirm the presence of a basiloid nest if above criteria is identified in the suspected carcinoma:

- No identifiable rete pegs in the area below the basiloid nest or dermal skin appendages.
- Blood vessel(s) directed towards the basiloid nest site.
- Presence of a hyporeflective periphery to the basiloid nest.
- Presence of hyperreflective stroma surrounding the basiloid nest.

Note that above criteria are not common to all nodular and superficial basal cell carcinomas sampled so far, and these criteria should always be used in conjunction with an assessment of the superficial presentation of the area scanned.

Morphology of squamous cell carcinoma in OCT images

The squamous cell carcinoma presents as an aberrant cell mound atypical to the surrounding tissue layers and standard skin layer morphology.

The squamous cell mound has no defined shape or form as seen in basiloid cell nests of basal cell carcinomas (Fig. 14). The scatter pattern of the squamous cell mound, however, is distinctly different in comparison to the surrounding tissue, but is not uniform throughout the carcinoma (Fig. 15).

OCT scan slices further through the squamous cell mass show distinct cyst-like bodies present within the squamous mound and evidence of keratinisation/keratin deposits in certain regions identified by the hyperreflective areas encountered.

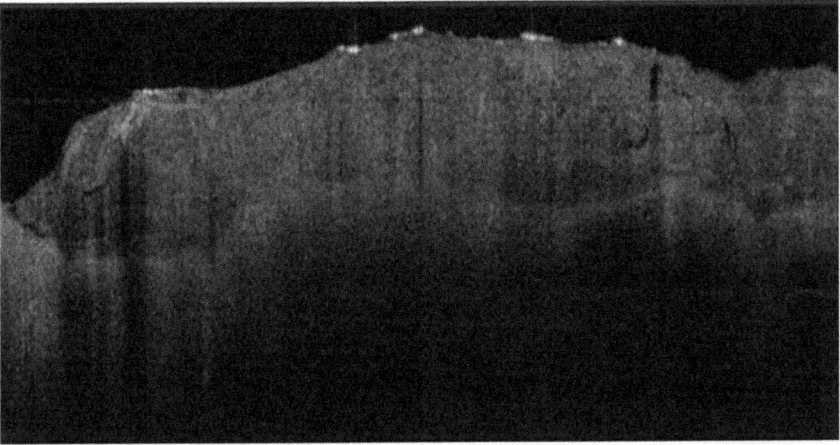

Fig. 14. OCT scan of a squamous cell carcinoma surface squamous cell mound.

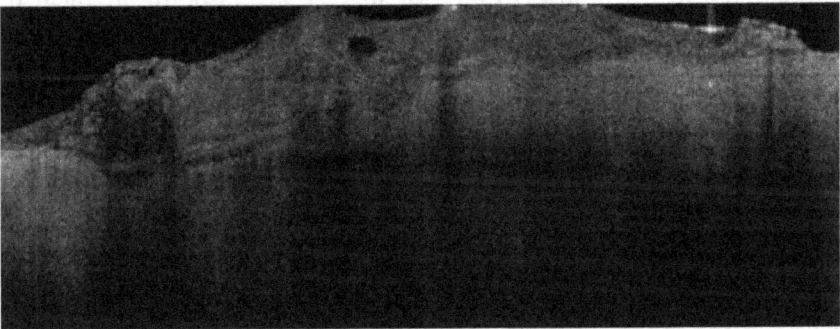

Fig. 15. OCT scan of squamous cell carcinoma (SCC) showing the surface squamous cell mound.

The afformentioned cyst-like bodies present within the mound appear to be void spaces but it is highly likely that they are fluid filled.

A distinct border between the base of the squamous cell mound and the disrupted Stratum Spinosum layer is visible; this may be due to high nuclear content, similar to that seen in the periphery of basal cell nests.

The squamous cell carcinoma can be identified by the expected surface squamous mound structure showing a clearly defined

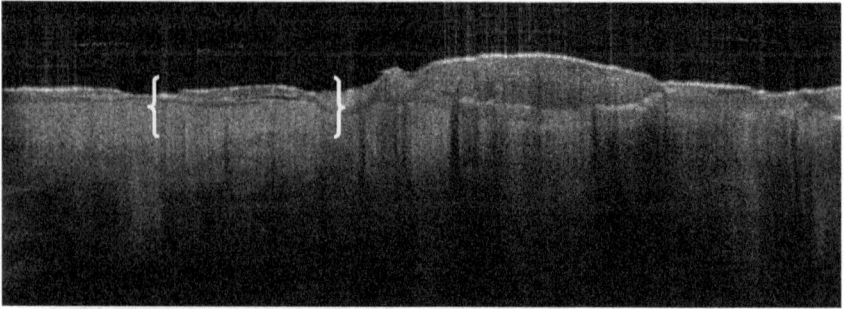

Fig. 16. OCT scan of a moderately differentiated squamous cell carcinoma region of surrounding surface layer abnormality indicated.

boundary with underlying tissue layers (Fig. 16). The carcinoma displays further skin surface abnormalities surrounding the squamous mass with the appearance of surface layers being atypical to healthy skin. The atypical surface layers resemble the appearance of minor surface ablation encountered in mechanically induced blisters however possessing an abnormal scatter pattern indicating the presence of keratin deposits within the protruding surface layer.

Squamous cell carcinomas and structures caused by the formation of squamous mounds are identifiable through optical coherence tomography. Scans of lesions suspected to be squamous cell carcinomas reveal structures which are atypical to the surrounding tissue layers of the epidermal skin regions.

Squamous cell carcinomas that present as protruding growths of squamous cell masses do not possess homogenous scatter pattern as is characteristic of basal cell carcinoma nests. The squamous mounds are primarily identified within the superficial layers of the epidermis mainly within the Stratum Spinosum and where applicable the Stratum Corneum.

Unlike basal cell carcinomas the structural appearance and form of the squamous cell carcinoma is not universal in all examples analysed.

OCT scans of some squamous cell carcinomas can identify distinct boundaries between the squamous mass and the surrounding tissue layers of the epidermis. This is identifiable not only through the difference in scatter pattern of the carcinoma tissue, but also due to the presence of a distinct boundary between the squamous mass and surrounding stromal cells. The boundary is usually more noticeable underlying the main focal point of the squamous mound. Similar to the basal cell carcinoma, the hyporeflective boundary may be a result of increased nuclear content resulting from an abnormally proliferating layer.

Akin to basal cell carcinomas the presence of a squamous cell carcinoma disrupts the traditional undulating pattern of the Stratum Spinosum layer and in all samples examined the interface between the Stratum Spinosum and the papillary dermis is affected.

The samples examined suggests that fluid filled cysts and cyst-like bodies form as a result of squamous mounds being present in the epidermis layers, many of which are located within the squamous mound but can also be found in tissue layers underlying the primary squamous mass.

Abnormal surface structures are also found in areas surrounding squamous mounds/masses. These structures appear akin to mechanically induced blisters in which the skin surface appears to be in a state of delamination, however in the case of those attributed to squamous cell carcinomas the partially delaminated layer is still connected to lower epidermal layers however shows increased scattering properties presumably due to keratinisation and the degree of detachment from the underlying skin layers.

Early clinical work has identified the following features that may be used to identify squamous cell carcinomas:

- Noticeable change in scatter pattern compared to surrounding tissue layers
- Distinct mound like structure protruding from established epidermal layers
- Presence of an identifiable border isolating the mound like structure from underlying tissue layers

The following features are not able to identify squamous cell carcinomas but rather are able to confirm the presence of an abnormal squamous mass if above criteria is identified in the suspected carcinoma:

- Presence of fluid filled cysts or cyst-like structures within or surrounding the squamous mound/mass
- Reduction or loss of skin appendages
- Evidence of keratin deposits and/or keratinisation in or surrounding the squamous mound

Like with basal cell carcinoma, it should be noted that above criteria are not common to all squamous cell carcinomas sampled so far. Criteria should always be used in conjunction with superficial presentation of the area scanned.

Clinical uses of OCT in skin cancer

OCT has a large number of potential uses in dermatology. It is what is commonly known as a "platform" technology. Significant opportunities exist to:

1. Improve surgical care
2. Allow non-surgical care
3. Allow monitoring of precursor lesions
4. Target biopsies more effectively
5. Optimize histological analysis
6. Allow diagnosis of disease

Of these, the author has been most involved in the mapping of surgical margins to improve surgical treatment of skin cancer, so this will be covered in most detail.

Improve surgical care

It is rare for a non-melanoma skin cancer to kill. Rather, the disease tends to be one that causes morbidity. Research shows that if clear histological margins are achieved then BCC is essentially cured — recurrence is only about 1% at 5 years.[19–21]

Unlike Melanoma, where most considerations come second to survival, in NMSC there is a balance to be struck between removal of sufficient tissue to affect a cure and preservation of as much tissue as possible. This preservation tends to be high on the list of patient concerns; much NMSC is caused by sun-damage, and as much as 80–85% of NMSC occurs on the head and neck, precisely where there is least "spare" tissue and where morbidity is of greatest concern.[22,23]

It is this balance between cure and harm that should be improved using OCT. Current standards of care are highest where Mohs micrographic surgery is used. Here, the obvious bulk tumor is surgically removed, and then the edges of the surgical defect are skimmed off, frozen, sectioned, mounted and examined microscopically while the patient waits. Any parts of the margin that are still involved are re-excised and the margins are checked again. In such a way, at the end of the day the patient leaves with guaranteed complete clearance of the lesion, while as much healthy tissue as possible is preserved.

While Mohs surgery offers an excellent result for the patient, it is costly, time consuming, and requires special facilities and support. Hence while it is extremely common in the USA, with specialized clinics offering the complete service under one roof, most other healthcare systems tend to make limited use of it. Many only use Mohs where preservation of function is particularly critical, or where lesions are particularly at risk of recurrence, for example for aggressive or recurrent lesions on the face.[23]

For most patients then, the standard of care is traditional surgical excision (SE). Here the clinician will make a judgment about the apparent margins of the lesion, normally based upon a combination of visual inspection, possibly using a dermatoscope, and on curettage of the bulk tumor. He or she will then make an excision, leaving a margin around the apparent borders.

Surgical excision is the optimal treatment for BCC.[19,21,22] If the tumor is completely removed then recurrence rates are low, varying from 0.3%–1.9%.[19–21]

These low rates of recurrence rely on complete excision of the clinical tumor in the first instance.

Studies have reported incomplete excision, where margins are involved with or extremely close to tumor, in 4.7%[19] to 7%[20] of cases reported in British Plastic Surgery units, and 6.3% in two retrospective studies in Australia.[24,25]

A German study[26] considered the success of surgical excision in a group of 199 pBCCs and 88 aggressive pBCCs. For pBCC, they reported incomplete excision after the first surgery in 18% of cases, with 13% of that group not clear after a second excision and thus requiring Mohs surgery. For aggressive pBCC 24% of tumors were incompletely excised during first surgery.

The problem, counter intuitively given this high failure rate, is not one of clinician skill or lack of care; clinicians tend to be highly skilled at identifying the extent of lesions at the surface of the skin. What causes problems is that the lesion may extend under the skin further than it does at the surface. With the need to preserve tissue for functional and aesthetic reasons high on the priority list, it is inevitable that very wide margins are not taken except when necessary.

OCT can help the clinician here by allowing them to visualize the extent of the lesion below the surface, allowing for a better balance between likelihood if incomplete resection and tissue removed. For conventional surgical excision this should mean better clearance rates. For Mohs surgery the clearance rate will not change: it is already as close to 100% as can be achieved. What will likely change is the number of stages, or cycles of surgery and histology, necessary to achieve clearance. Currently this stands at an average of 1.76.[11]

A practical implementation of margin mapping

To test this margin-mapping hypothesis, the author has been undertaking a series of studies in conjunction with Prof Daniel Siegel and his colleagues at Long Island Skin Cancer, Smithtown, NY, USA. LISC is a private clinic seeing upwards of 40 patients a week for Mohs surgery on biopsy-confirmed skin cancers.

During these IRB approved trials, a protocol was developed to test the hypothesis that the OCT system, in this case a Michelson

Diagnostics VivoSight Multi-beam system, could visualize the margins of NMSC lesions. A conflict of interest should be declared here — the author is a founder and executive director of Michelson Diagnostics.

The pilot study aimed to correlate the Mohs defect, the margin at which the lesion has been demonstrated to be histologically clear, with the margins indicated using the OCT system. The study presented two correlation challenges: co-correlation to photographs from before, during and after the Mohs procedure, and accurately translating the margin visualized on the OCT system to a point on the patient's skin. For the latter, what was needed was a reference point between the two. To achieve this, an adhesive paper marker was affixed to the skin, centered on the clinically apparent lesional margin. The OCT instrument was then used to take a series of fifty 5 mm wide × 2 mm deep cross-sectional images of the region enclosed by the marker, with each image separated by 0.1 mm so that a total area of 5 × 5 mm was evenly sampled. It was typically possible to image to a depth of about 1.25 mm into the skin, although there was surprising variation in this figure from patient to patient. This variation could partially be attributed to variations in the thickness of the highly reflecting keratinized top layer of the skin, but some patients' skin simply seems more "dense" than others.

The images were then viewed, and based on visualization of morphological features within the sample, the apparent transition point between tumor and benign perilesional skin was identified. This judgment was aided by reference to a "normal" — skin sample imaged contra laterally to the lesion where possible or simply nearby if not, which was imaged first to serve as a frame of reference for evaluating the patient's local skin morphology.

By reference to the imaging frame number and the reference number, the frame containing the identified transition point could be transposed to a point on the patient's skin, and a mark was made to denote the predicted margin. This procedure was repeated an additional three times, so that the superior, inferior, lateral, and medial lesion boundaries were delineated.

The Mohs procedure was then performed in the usual fashion, with no reference to the indicated borders determined by OCT analysis. Once the margins were found to be clear of tumor, the accuracy of the OCT margin could be compared to that achieved by the Mohs procedure.

At the time of writing (October 2010), the initial studies have been sufficiently promising that a full study is in preparation to demonstrate efficacy of this method. These initial results will be published shortly.

Increased use of non-surgical therapy

There is good evidence that many superficial or low-risk NMSC lesions respond extremely well to non-surgical treatments such as Photodynamic Therapy (PDT) or topical chemotherapy.

PDT is a technique in which the patient is given a photo-sensitive drug that targets tumor cells, and is then exposed to light which causes the tumor cells to be destroyed. It has the advantage that cosmetic damage is minimized. A course of treatments is required and there is no effective way of determining whether the treatment is working — if not, the cancer re-grows. It is more effective for superficial cancers.

PDT is not widely available in the USA because the photosensitive drug cleared by the FDA is not yet explicitly approved for most uses in dermatology. 20% 5-aminolevulinic acid (ALA) is approved for the treatment of nonhyperkeratotic actinic keratoses (AKs) of the face and scalp.[27] For PDT, porphyrin (5-aminolevulinic acid [ALA], methyl aminolevulinate) is generally applied topically to the lesion using an applicator stick. Although the porphyrin can be administered systemically, this approach is used rarely since systemic therapy can be associated with prolonged photosensitivity.

Several hours after topical application of the porphyrin, the area is exposed to visible light in the 630 to 635 nanometer range. The light penetrates the first 5 to 6 mm of skin and is selectively absorbed by the photosensitizer, which generates reactive oxygen species. This in turn can cause lipid peroxidation, protein crosslinking, increased membrane permeability, and ultimately cell death. In addition to direct effects on tumor cells, PDT can damage blood vessels, resulting in impaired blood flow, and can stimulate a vigorous local inflammatory reaction that may contribute to tumor destruction.

All other entities treated, and any variation from standard parameters, are considered off-label in the eyes of the FDA, although documented clinical trials do support such alternate uses. So PDT is generally only used in cases where it can be justified (and therefore reimbursed) by specific circumstance. PDT is more widely used in Europe, where Phoscan, rather than ALA, is normally used.

Topical chemotherapy

In the USA, the most commonly used topical chemotherapy drug is 5-Fluorouracil (5-FU). 5-FU inhibits thymidylate synthetase, disrupting DNA synthesis. The literature suggests that 5-FU should be restricted to superficial BCCs in noncritical locations. Treatment of non-superficial, recurrent, and other high-risk BCCs with 5-FU results in low cure rates. Furthermore, topical treatment of lesions that are not superficial can give the false impression of a cure despite persistent dermal disease deeper in the tissue. For this reason, deep, nodular or high-risk BCCs are generally considered contraindications to topical 5-FU.

Another topical therapy becoming available is Imiquimod, a novel immune response modifier that is FDA-approved for the treatment of superficial BCCs in low-risk sites. Imiquimod is thought to promote apoptosis in skin cancer cells by circumventing the tumor cells' anti-apoptotic mechanisms,[28] and/or by stimulating monocytes/macrophages and dendritic cells to produce cytokines that stimulate cell-mediated immunity.[29]

Monitor precursor lesions

It has already been mentioned that most NMSC lesions are a result of sun damage. Because of this, it is unusual for a NMSC lesion to occur in otherwise completely normal skin — it is normally only the focus of a more general field change in the skin. As a part of this, pre-cancerous lesions are a common occurrence. Actinic

Keratoses are a precursor to SCC, and Bowens Disease a early/mild form of BCC. There is some debate about the best way to treat these precursor lesions. It is not uncommon for them to regress as well as progress, for example in patients who start to make good use of sunscreen, so monitoring is often chosen instead of early treatment. OCT has the potential to improve this process, as it allows the clinician to inspect more features of the lesions than are clinically apparent. A lack of invasion, for example, can be checked without recourse to a biopsy. The clinician could also check for factors such as changes in thickness of the lesions over time, indicating change and therefore a possible need for active intervention.

In this way, OCT offers the possibility of treatment more tailored to the behavior of these pre-malignant lesions, avoiding unnecessary surgery without risking unwanted progression.

Allow diagnosis of disease

There have been several recent studies that have attempted to investigate the use of OCT as a diagnostic tool for NMSC in a systematic and structured manner. Groups in Denmark and Germany have used conventional single beam systems to image patients in large, blinded studies and have attempted to determine biomarkers and other indicators of disease.

Mette Mogensen aimed to describe OCT features in NMSC such as actinic keratosis (AK) and basal cell carcinoma (BCC) and in benign lesions and to assess the diagnostic accuracy of OCT in differentiating NMSC from benign lesions and normal skin.[30] She took OCT and polarization-sensitive (PS) OCT from 104 patients, and performed observer-blinded evaluation of OCT images from 64 BCCs, one baso-squamous carcinoma, 39 AKs, two malignant melanomas, nine benign lesions, and 105 OCT images from perilesional skin. 50 OCT images of NMSC and 50 PS-OCT images of normal skin were evaluated twice.

Sensitivity was 79% to 94% and specificity 85% to 96% in differentiating normal skin from lesions. Important features were absence

of well-defined layering in OCT and PS-OCT images and dark lobules in BCC. Discrimination of AK from BCC had an error rate of
50% to 52%.

Mogensen reported:

"A break-up of the characteristic layering in normal skin is found
in OCT images of NMSC and malignant melanoma (MM) lesions,
but this break-up is also seen in various benign lesions such as seb-
orrheic keratosis and benign melanocytic nevi. We chose to have
observers examine five key OCT image features selected from earlier studies and according to our own experience:

- disruption of layering (demonstrated in AK and BCC);
- white streaks and dots in epidermis (hyperkeratosis in AK);
- other focal changes including thickening of the epidermis (AK);
- dark rounded areas, sometimes surrounded by a white area
  (BCC basaloid island cell clusters and surrounding stroma);
- presence of a nonbirefringent, homogeneous band in the
  upper part of PS-OCT images, which is approximately 200 m in
  normal images corresponding to epidermis and papillary dermis. The change in collagen architecture identifies the border
  between papillary dermis and reticular dermis in PS-OCT
  images from normal skin. Absence or break-up of this band has
  been described in invasive BCC".

Melanoma

The early recognition and therapy of skin tumors, especially of the
early and precursor lesions, can increase the survival of patients
significantly, and can reduce the huge social-economic damage
currently experienced. Currently the early recognition of tumors of
the skin relies solely on clinical inspection, on epiluminescence or
confocal microscopy, and on subsequent biopsy, because the common imaging systems such as ultrasound, CT and MRT do not have
sufficient sensitivity for the recognition of early forms of skin cancer.
The current procedures are expensive and time consuming.

Additionally, single biopsies may not lead to the correct diagnosis, because lesions commonly occur in multiple locations, and they can also occur over large regions before they become invasive at selected areas. Therefore the crucial areas of invasive tumor growth may not be identified by single biopsies.

Target biopsies more effectively

A biopsy is the definitive diagnostic approach at present. Use of fixed section histology, with the attendant stains and molecular tests, can give a diagnosis with an extremely high degree of confidence. The problem is, that single biopsies may not lead to the correct diagnosis as lesions commonly occur in multiple locations, and they can also occur over large regions before they become invasive at selected areas. If that sample has not been taken from the most representative point of the lesion, the crucial areas of invasive tumor growth may not be identified by single biopsies.

OCT offers the clinician a means to visualize the lesion before they take the biopsy. In this way, they can visualize the points at which the tissue looks most worrying, whether that manifests as an absence of the DEJ, obvious cell nests, or abnormal vasculature, meaning that when the biopsy is taken, it will give the most relevant result.

In the short term, this has the potential to improve current standards of care. In the medium term, as more information is gathered about the diagnostic potential of OCT, it may even allow some biopsies to be avoided.

Optimize histological analysis

The staging of some skin tumors, notably melanoma, relies heavily on the maximum depth of the tumor. If a melanoma is less than 1 mm thick then the prognosis, and therefore treatment, is very different to that for a thicker lesion. For example, there would be a theoretical ability, based on tumour thickness, to predict the need for neck dissection to remove metastatic nodes in melanoma of the head and neck skin. Studies have indicated that if a lesion is less than 0.75 mm then this would not be necessary.[31]

The problem in judging tumor depth is typically is one of sampling density. Normally the lesion will be excised, and then a number of sections will be made in the manner of a sliced loaf. The depth of the lesion will then be established and reported based on the maximum depth visualized. The problem is that these sections are normally quite far apart on a histological scale — perhaps 1 mm. As a result, it is suspected by some that there may be lesions in which deeper projections escape notice as they fall between these slices and are missed.

OCT could be used *ex vivo* to assist the dermatopathologist in deciding where to make their sections. By identifying the deepest points of the lesion before sections are made, the histology sections can be assured of including the most critical points.

Summary

OCT has progressed enormously since the early '90s when it was first conceived. Academic, commercial and clinical groups worldwide are steadily addressing the twin barriers of technical capability and clinical understanding.

The potential benefit for all is clear. More accurate, more timely, more cost effective and more effective treatment of skin cancer would be made possible by giving the clinical dermatologist the ability to see into the skin. That point is rapidly becoming a reality as understanding grows. As for the future, it seems likely that OCT will continue to develop. Ever better images at higher resolution, automated diagnosis, additional functional measurements such as blood flow analysis and polarization sensitivity may add further diagnostic markers.

Further in the future, OCT may be combined with other imaging and diagnostic modalities to give a comprehensive diagnostic tool. For example, in combination with confocal microscopy or hyperspectral imaging, the diagnosis and staging of a superficial melanoma could be done entirely non-invasively.

References

1. Tanno N, Ichikawa T, Saeki A. (1990) *Lightwave Reflection Measurement*, Japan.
2. Huang D, Swanson E, Lin C *et al.* (1991) Optical coherence tomography. *Science* **254**: 1178–1181.
3. Fercher AF, Hitzenberger CK, Kamp G, El-Zaiat SY. (1995) Measurement of intraocular distances by backscattering spectral interferometry. *Optics Communications* **117**: 43–48.
4. Fercher AF, Drexler W, Hitzenberger CK, Kamp G. (1994) Measurement of optical distances by optical spectrum modulation. Fercher AF *et al.* (ed.), Vol. 2083, pp. 263–267. Presented at Microscopy, Holography, and Interferometry in Biomedicine, SPIE, Budapest, Hungary.
5. Holmes J. (2008) Theory and applications of multi-beam OCT. Podoleanu A. (ed.), Vol. 7139, pp. 713908–7. Presented at 1st Canterbury Workshop on Optical Coherence Tomography and Adaptive Optics, SPIE, Canterbury, United Kingdom.
6. Tomlins PH, Woolliams P, Tedaldi M *et al.* (2008) Measurement of the three-dimensional point-spread function in an optical coherence tomography imaging system. *Proc. SPIE* **6847**: 68472Q.
7. Marmur ES, Berkowitz EZ, Fuchs BS *et al.* (2010) Use of high-frequency, high-resolution ultrasound before Mohs surgery. *Dermatol Surg* **36**(6): 841–847.
8. Bialy TL, Whalen J, Veledar E *et al.* (2004) Mohs micrographic surgery vs traditional surgical excision: A cost comparison analysis. *Arch Dermatol* **140**(6): 736–742.
9. Moore JV, Allan E. (2003) Pulsed ultrasound measurements of depth and regression of basal cell carcinomas after photodynamic therapy: Relationship to probability of 1-year local control. *Br J Dermatol* **149**(5): 1035–1040.
10. Mogensen M, Nurnberg BM, Forman JL *et al.* (2009) *In vivo* thickness measurement of basal cell carcinoma and actinic keratosis with optical coherence tomography and 20-MHz ultrasound. *Br J Dermatol* **160**(5): 1026–1033.
11. Rogers HW, Weinstock MA, Harris AR *et al.* (2010) Incidence estimate of nonmelanoma skin cancer in the United States, 2006. *Arch Dermatol* **146**(3): 283–287.

12. Steiner RW, Kunzi-Rapp K. (2006) Optical coherence tomography versus ultrasound for diagnostics in dermatology. Frenz M *et al.* (ed.), Vol. 6284, pp. 62840E. Presented at International Conference on Lasers, Applications, and Technologies 2005: Laser Technologies for Environmental Monitoring and Ecological Applications, and Laser Technologies for Medicine, SPIE.

13. Mogensen M, Thrane L, Jorgensen TM, Jemec GBE. (2007) Diagnostic potential of optical coherence tomography in non-melanoma skin cancer: A clinical study. Andersen PE, Chen Z. (ed.), Vol. 6627, pp. 662708-8. Presented at Optical Coherence Tomography and Coherence Techniques III, SPIE, Munich, Germany.

14. Gambichler T, Orlikov A, Vasa R *et al.* (2007) *In vivo* optical coherence tomography of basal cell carcinoma. *J Dermatol Sci* **45**(3): 167–173.

15. Pfau PR, Sivak MV, Jr., Chak A *et al.* (2003) Criteria for the diagnosis of dysplasia by endoscopic optical coherence tomography. *Gastrointest Endosc* **58**(2): 196–202.

16. Gambichler T, Moussa G, Sand M *et al.* (2005) Correlation between clinical scoring of allergic patch test reactions and optical coherence tomography. *J Biomed Opt* **10**(6): 064030.

17. Olmedo JM, Warschaw KE, Schmitt JM, Swanson DL. (2006) Optical coherence tomography for the characterization of basal cell carcinoma *in vivo*: A pilot study. *J Am Acad Dermatol* **55**(3): 408–412.

18. Olmedo JM, Warschaw KE, Schmitt JM Swanson DL. (2007) Correlation of thickness of basal cell carcinoma by optical coherence tomography *in vivo* and routine histologic findings: A pilot study. *Dermatol Surg* **33**(4): 421–5; discussion 425–426.

19. Kumar P, Orton CI, McWilliam LJ, Watson S. (2000) Incidence of incomplete excision in surgically treated basal cell carcinoma: A retrospective clinical audit. *Br J Plast Surg* **53**(7): 563–566.

20. Griffiths RW. (1999) Audit of histologically incompletely excised basal cell carcinomas: Recommendations for management by re-excision. *Br J Plast Surg* **52**(1): 24–28.

21. Griffiths RW, Suvarna SK, Stone J. (2005) Do basal cell carcinomas recur after complete conventional surgical excision? *Br J Plast Surg* **58**(6): 795–805.

22. Griffiths RW, Suvarna SK, Stone J. (2007) Basal cell carcinoma histological clearance margins: An analysis of 1539 conventionally excised tumours. Wider still and deeper? *J Plast Reconstr Aesthet Surg* **60**(1): 41–47.

23. Telfer NR, Colver GB, Morton CA (2008) Guidelines for the management of basal cell carcinoma. *Br J Dermatol* **159**(1): 35–48.

24. Dieu T, Macleod AM. (2002) Incomplete excision of basal cell carcinomas: A retrospective audit. *ANZ J Surg* **72**(3): 219–221.

25. Sussman LA, Liggins DF. (1996) Incompletely excised basal cell carcinoma: A management dilemma? *Aust N Z J Surg* **66**(5): 276–278.

26. Mosterd K, Arits AH, Thissen MR, Kelleners-Smeets NW. (2009) Histology-based treatment of basal cell carcinoma. *Acta Derm Venereol* **89**(5): 454–458.

27. Gold MH. (2007) Photodynamic therapy update 2007. *J Drugs Dermatol* **6**(11): 1131–1137.

28. Schon M, Bong AB, Drewniok C *et al.* (2003) Tumor-selective induction of apoptosis and the small-molecule immune response modifier imiquimod. *J Natl Cancer Inst* **95**(15): 1138–1149.

29. Urosevic M, Dummer R, Conrad C *et al.* (2005) Disease-independent skin recruitment and activation of plasmacytoid predendritic cells following imiquimod treatment. *J Natl Cancer Inst* **97**(15): 1143–1153.

30. Mogensen M, Joergensen TM, Nurnberg BM *et al.* (2009) Assessment of optical coherence tomography imaging in the diagnosis of non-melanoma skin cancer and benign lesions versus normal skin: Observer-blinded evaluation by dermatologists and pathologists. *Dermatol Surg* **35**(6): 965–972.

31. Cecchi R, Buralli L, Innocenti S, De Gaudio C. (2007) Sentinel lymph node biopsy in patients with thin melanomas. *J Dermatol* **34**(8): 512–515.

Index

www.ingramcontent.com/pod-product-compliance
Lightning Source LLC
Chambersburg PA
CBHW050603190326
41458CB00007B/2154

* 9 7 8 9 8 1 4 2 9 5 4 0 6 *